THE HERBERT SPENCER LECTURES 1986

Evolution and its Influence

Evolution and its Influence

Edited by
ALAN GRAFEN

CLARENDON PRESS · OXFORD
1989

Oxford University Press, Walton Street, Oxford OX2 6DP

Oxford New York Toronto
Delhi Bombay Calcutta Madras Karachi
Petaling Jaya Singapore Hong Kong Tokyo
Nairobi Dar es Salaam Cape Town
Melbourne Auckland

and associated companies in
Berlin Ibadan

Oxford is a trade mark of Oxford University Press

Published in the United States
by Oxford University Press, New York

British Library Cataloguing in Publication Data

Evolution and its influence. — (The Herbert
Spencer lectures; 1986)
1. Intellectual life. Influence of theories
of evolution of Darwin, Charles, 1809–
1882. 1800
I. Grafen, Alan II. Series
909.8
ISBN 0–19–827275–8

Library of Congress Cataloging in Publication Data

Evolution and its influence/edited by Alan Grafen.
(The Herbert Spencer lectures; 1986)
Includes index.
1. Evolution. I. Grafen, Alan. II. Series.
QH371.E92 1989   575—dc19   88–29748
ISBN 0–19–827275–8

Set by Cambrian Typesetters
Printed in Great Britain by Biddles Ltd
Guildford and King's Lynn

Preface

Professor Henry Harris introduced the 1986 series of Herbert Spencer Lectures with an explanation of its purpose. The series was to survey the influence on various disciplines of 'the greatest intellectual revolution of the nineteenth century', the Darwinian view of evolution by natural selection.

Professor A. J. Cain begins with a historical summary of ideas of evolution and natural selection, and an account of the enormously important contribution of Charles Darwin. A discussion of the evidence that now decisively favours Darwin's views is followed by an explanation of the fallacy in the popular argument that natural selection is a tautology. Professor Cain then considers what the fact that man has evolved might suggest to us about his mental capacities.

W. G. Runciman traces the fates of Spencer's and Marx's ideas of evolution in sociology, and identifies the same flaw in both, namely the idea that evolution is assumed to have an intrinsic purpose or direction. Darwin's revolution was not the hypothesizing of a new or primitive or vital force, but the realization of the far-reaching consequences of familiar facts. Mr Runciman reports that his research efforts are currently directed to evolution in sociology without an idea of progress, but with the idea that there are regularities, including selective processes, at work in the behaviour of the components of society, which will allow the diversity and design of societies to be understood as their unintended consequences.

Professor R. J. Herrnstein's work is in psychology and concerns the causation of behaviour. This roughly means trying to predict what a human or other animal will do. He reports experiments which show that humans perform irrationally on simple tasks, and that hungry pigeons fail to maximize their rate of feeding from simple apparatus. Pigeons and humans behave in the same suboptimal way in these two similar experiments, which provide a strong confirmation of Professor Herrnstein's preferred model of the control of behaviour. He makes an analogy between evolution as the resultant of a conflict between individuals, which may not be for the general advantage of the species, and the behaviour of an individual as the resultant of changes in the forces that operate in his model, which may not be to the best advantage of the individual.

Central issues in social anthropology are confronted by Professor Maurice Godelier in his discussion of the evolution of man's ideas of kinship and of incest prohibition. He expresses Lévi-Strauss's fundamental step forward as the realization that human beings acted deliberately to bring order to society by inventing relations of kinship. Professor Godelier then argues that while the prohibition of incest is indeed an adaptive response, relations of kinship themselves should 'be seen, not as the goal pursued by the institution of this sexual taboo, but as its quasi-automatic, and indeed almost involuntary outcome'.

Professor Bernard Williams discusses three themes as instances of evolutionary thought in philosophy. He points to a dangerous vacuity in an analogy of Sir Karl Popper between natural selection and scientific progress. He then discusses evolutionary epistemology, which deduces from the fact that man has evolved what his knowledge is like, with qualified approval. Professor Williams then describes the consequences of the fact of evolution for ethics, arguing that most of its apparent consequences do not follow.

Many of the forces that mould a biological species arise outside the species, and the relevant parts of the environment of a species are called its niche. Sir Ernst Gombrich's theme is that many of the forces involved in changes in art forms are not intrinsic to art, and that by analogy an art form may usefully be considered to have a niche. The effects of liturgical changes on art are a case in point: when the officiating priest ceased to face the congregation from behind the altar but changed his position and orientation, the altar could be adorned with images. The ambition of donors and the aspirations of artists subsequently led to an increasing emancipation of these images from their religious function and to the development of secular easel painting.

These are the six Herbert Spencer Lectures for 1986. The aim of the series will have been achieved if the reader gains an impression of evolution's influence on the wide range of disciplines represented here.

A.G.

Zoology Department, Oxford University
September 1987

Contents

List of Plates

Notes On Contributors

A. J. CAIN is Derby Professor of Zoology at Liverpool University. While a Lecturer in Animal Taxonomy at Oxford he led expeditions to Melanesia and South America to study evolution, especially in birds. He has published numerous scientific papers, chiefly on aspects of evolution and on the theory and history of classification. He is a Foreign Fellow of the Academy of Natural Sciences, Philadelphia, and President of the European Society for Evolutionary Biology.

MAURICE GODELIER is Professor at the École des Hautes Études en Sciences Sociales, Paris and was Scientific Director of the French National Centre for Scientific Research from 1982–6. Since 1960 he has held a number of positions, working with, amongst others, Fernand Braudel and Claude Lévi-Strauss, and has undertaken extended periods of field-work among the Baruya of the Highlands of Papua New Guinea. His publications include *Rationality and Irrationality in Economics* (trans. 1972), *Marxist Perspectives in Anthropology* (trans. 1977), *The Making of Great Men* (trans. 1986), and *The Mental and the Material* (trans. 1986).

Sir ERNST GOMBRICH was Director of the Warburg Institute and Professor of the History of the Classical Tradition at London University from 1959–76. He has been Slade Professor of Fine Art in the Universities of Oxford (1950–3) and of Cambridge (1951–63) and is a Fellow of the British Academy and member of many other learned bodies. He has been awarded the Order of Merit and is the recipient, *inter alia*, of the Erasmus Prize and the International Balzan Prize. His many publications include *The Story of Art* (1950), *Art and Illusion* (1960), *Aby Warburg* (1970), and *The Sense of Order* (1979).

RICHARD J. HERRNSTEIN is Edgar Pierce Professor of Psychology at Harvard University and a member of the American Academy of Arts and Sciences. His work has been concerned with the application of mathematical models to learning theory. His book *I.Q. in the Meritocracy* (1973) considers the respective roles of heredity and environment in the formation of intelligence. He is also the co-author of *A Source Book in the History of Psychology* (1965) and *Psychology* (1975).

W. G. RUNCIMAN is Senior Research Fellow of Trinity College, Cambridge, a Fellow of the British Academy, and a Foreign Honorary

Member of the American Academy of Arts and Sciences. His publications include *Relative Deprivation and Social Justice* (1966), *A Critique of Max Weber's Philosophy of Social Science* (1972), and *A Treatise on Social Theory*: vol. i (1983), vol. ii (1988).

BERNARD WILLIAMS is Professor of Philosophy at the University of California, Berkeley. He was Provost of King's College, Cambridge from 1979–87 and has held chairs at the Universities of Cambridge and London and visiting professorships or fellowships at Princeton, Harvard, Berkeley, and the Australian National University. Professor Williams is a Fellow of the British Academy, a Foreign Honorary Member of the American Academy of Arts and Sciences, and was Chairman of the Committee on Obscenity and Film Censorship. He is also a Director of the English National Opera. His publications include *Problems of the Self* (1973), *Descartes: The Project of Pure Enquiry* (1978), and *Ethics and the Limits of Philosophy* (1985).

1

The True Meaning of Darwinian Evolution[1]

A. J. CAIN

Introduction

Herbert Spencer was a thorough-going evolutionist long before the *Origin of Species*, and his ideas on evolution deserve a more sympathetic examination than they have had for nearly a century. For example, his inference that the oxygen content of the atmosphere must have increased by biological action and made possible the evolution of warm-blooded vertebrates was a remarkable one for its date (1844), anticipating some very recent work on the evolution of the biosphere.[2] Later,[3] he accepted Charles Darwin's work, agreeing that many phenomena such as mimicry could be produced only by natural selection, but insisting that many others could not; only a Lamarckian form of evolution (by use and disuse) could have produced them. As he said himself in the *Autobiography*, he came close in some of his socio-political writings to natural selection in Darwin's sense but missed his own implications.

Charles Darwin

Spencer's publications, like those of Lamarck, Erasmus Darwin, and Robert Chambers (to mention only three), are historic

[1] I am very grateful for criticisms received from: A. F. Brown, B. C. Clarke, G. M. Davis, R. Dawkins, R. I. M. Dunbar, J. S. Jones, E. Mayr, G. A. Parker, R. G. Pearson, M. Stanisstreet, T. E. Thompson, and B. Wood.

[2] H. Spencer, 'Remarks on the Theory of Reciprocal Dependence in the Animal and Vegetable Creations, as Regards its Bearing upon Palaeontology', *An Autobiography*, vol. i (London, 1844; repr. 1904). App. F.

[3] Id., *Factors of Organic Evolution* (London, 1887); id., *The Inadequacy of Natural Selection* (London, 1893); id., *A Rejoinder to Professor Weismann* (London, 1893).

evidence that Charles Darwin did not invent the idea of evolution.
Nor did he invent natural selection in its stabilizing action; it had
been known for two thousand years that monsters, weaklings, and
misfits tended to perish early. Nor did he invent sexual selection.
His grandfather Erasmus Darwin had done that and given a
correct example, stags fighting for does. The horns of the stag 'are
sharp to offend his adversary, but are branched for the purpose of
parrying or receiving the thrusts of horns similar to his own, and
have therefore been formed for the purpose of combating other
stags for the exclusive possession of the females; who are
observed, like the ladies in the times of chivalry, to attend the car
of the victor.' Male birds (cocks and quails) fight for the same
reason, and 'The final cause of this contest amongst the males
seems to be, that the strongest and most active animal should
propagate the species, which should thence become improved.'[4]

A hasty reader might imagine that Erasmus Darwin was
therefore the originator of evolution by natural selection. It is true
that he was as aware of artificial selection as was his grandson; but
although Robert Bakewell, the Earl of Leicester, and others were
achieving miracles of improvement in their breeds of domestic
animals, not one of them even hinted at the possibility of
improving a cow into a different species and so on indefinitely.
Erasmus Darwin's theory of evolution, as became a physician, was
an embryological, teratological, psychological one that wellnigh
out-Lamarcked Lamarck. (Samuel Butler's use of it to castigate
Charles Darwin was thoroughly unscrupulous.)

As it turned out later, Charles Darwin was not even the first to
think of natural selection as producing *indefinite change* in living
things. But neither of his predecessors, William Wells and Patrick
Mathew, saw beyond their immediate problems and observations.
They never realized that they had found the key to the explanation
of how the vast diversity of beings had been produced, and to their
(often exquisite) adaptations. His distinctive contributions were
threefold:

(1) He realized that evolution by natural selection would explain
    brilliantly the facts of adaptive radiation and biogeography, of
    classification in a hierarchy of ranks, of vestigial organs and

---

[4] Erasmus Darwin, *Zoonomia; or, The Laws of Organic Life*, vol. i (London, 1794),
p. 503.

the strange courses of embryological development, and he documented its successes carefully.

(2) He realized from the facts of inheritance and over-production that there must be such a process as natural selection, acting as does artificial selection. And this latter had already produced the tail of the fantail pigeon, a character unique not only in the species but in the genus, family, order, and indeed the whole Class of birds; variation, therefore, could exceed by far specific limits.

(3) He admitted fully the apparent difficulties in the way of this theory, documented them just as carefully, and showed that they were probably not substantial, with a candour totally lacking in any fundamentalist publication against evolution that I have ever seen.

From then on, as T. H. Huxley realized,[5] evolution by natural selection was a scientific theory, open to verification by directed observation and experiment. A supernatural theory is not, because any difficulty can always be explained as so ordained by God.

Present Standing of Darwinian Evolution

The main evidence now for evolution by natural selection is threefold, from biogeography, palaeontology, and the genetics of wild populations. Biogeography shows us different animals (and plants) in different regions of the world becoming adapted for the same modes of life, and converging, in consequence, in their characteristics. In South America, for example (which, as plate tectonics shows on wholly independent evidence, has been an island continent like Australia from seventy million years ago until very recently) there have been rivers, lakes, mountains, and plains, full of plants, snails, and insects, all available to those birds and mammals that could get there to exploit them. In Guyana there is a broad-billed flycatcher, like those of other tropical regions in its bill, feet, and wing shape. There is a shrike with a black mask over its eyes and a hook on the end of its bill. There is even a pied

---

[5] T. H. Huxley, 'The Darwinian Hypothesis' (26 Dec. 1859), repr. in *Darwiniana: Essays*, vol. ii (London, 1893); id., 'Time and Life: Mr. Darwin's *Origin of Species*', *Macmillan's Magazine*, 1 (Dec. 1859), 142–8.

wagtail, with a large hind toe suitable for a bird that runs on the ground, a short stubby bill for snapping up insects, and a tail that wags up and down. Yet their anatomy shows that these birds are not what they seem; they are related to one another, and belong to a somewhat primitive group of perching birds which have had the ecological and evolutionary opportunities in South America that flycatchers, shrikes, and wagtails have had in Africa, and yet other unrelated stocks in Australasia.

This is not an isolated example. Such local adaptive radiation is seen in every group of plants and animals that has not had the dispersal facilities to get to every place in the world where they can live. Such exceptions include some rotifers and nematodes which can be dried to dust and blown around the world. Insects also, as such, have colonized everywhere as has Man. We live on too small a planet; were we on one the size of Jupiter, we might see different groups producing insects, birds, or Man independently in suffi-ciently isolated regions. We do see in the fossil record horses, defined as browsing or grazing quadrupeds escaping from their predators by running on a single digit per foot, evolving three times—the true horses (i.e. those we met first—there's nothing especially true about them) in North America, the South American litopterns derived from a similar primitive stock but independently, and in Africa the Miocene pseudhippomorphs derived wholly independently from the same stock as the conies of the Bible. No one can appreciate fully the pervasiveness of adaptive radiation and convergence who has not made a detailed study of the taxonomy and biogeography of a large group of living things; yet very few undergraduates in biology have the opportunity to do so. Except incidentally, taxonomy and biogeography have vanished from British universities, and since they do not produce immediate technological benefits they may soon be gone altogether. More-over, because of the rate at which the diversity of terrestrial life, at least, is becoming extinct, the material for such a study will soon be unavailable.

Evolution is completely *ad hoc*. It is what organism happens to be there at the right time and can take it that gets a new ecological job, not what is already doing it in some distant region, nor what is perfectly adapted for it. It was a tiny local radiation of birds in the Galapagos Islands that finally convinced Charles Darwin that evolution had occurred; in it could be seen in miniature what we now know has occurred everywhere.

Moreover, we can relate the fossil record of adaptive radiation to the independent evidence of continental drift. The greater reptiles evolved when most of the dry land of the continents was a single continuous mass. They spread all over it, and never achieved the diversity of the major groups of mammals. These came later, when the present continents were already mostly separated, and radiated on each of them so that the ant-eating habit, for example, could be evolved independently in South America (true ant-eaters), Africa (aardvarks), South-east Asia (pangolins), and Australia (the marsupial ant-eater *Myrmecobius* and the egg-laying echidnas). The distortions and misrepresentations (to put it no more strongly) of fundamentalists in regard to the fossil record, geology, and evolution are well documented by Ruse.[6]

Natural Selection

It has been (and still is) constantly objected against Darwinian evolution that it rests on a tautology. Natural selection means the survival of the fittest, but we know the fittest because they are the ones that survive. This was an apparently valid argument for some biologists and philosophers who sat around and pontificated about evolution, for or against, without ever looking seriously at any living organism in the wild. In part they did so because it was believed that evolution took too long for the detection of natural selection in action to be possible. The discovery of geological time was one of the most impressive scientific discoveries of the nineteenth century. And in part this attitude was reinforced by the grammatical structure of the English language which allows one to talk easily about things being adapted (or important) with no compulsion to state *for what*. (To misuse the terminology, these adjectives are often intransitive, but should be transitive always.)

To say that one organism is fitter than another means that, in its time and place, it has carried out some function (hiding itself, holding its breath while diving deeply, landing upside down on the ceiling, what you will) more efficiently than did the other

[6] M. Ruse, *Darwinism Defended* . . . (London, 1982).

individual. If this resulted in it having more descendants in the next generation than did the other individual, this fitness was producing an evolutionary consequence, the spread of the genes mediating the increased fitness. When the function is identified and analysed, and the genetic basis determined, there is nothing whatever tautologous about natural selection; moreover, on the basis of this knowledge one can predict which individual or genetic types in a given batch will be more successful. Nor is there tautology when a systematic relationship between gene frequencies and definite features of the environment can be demonstrated, even if the causes of this relationship are unknown. A strong relationship between increased resistance to an insecticide and length of time for which it has been applied (to take an all-too-common example of present-day evolution) is evidence of selection, even if we are wholly ignorant about how the insect does it. *Systematic* change can only be produced by natural selection. Fortunately, Endler's book on natural selection in the wild[7] gives an excellent survey of this topic with a lot of concrete examples.

Much of the earlier work on natural selection in the wild was done in Britain and a large part of it in Oxford. Lewontin[8] has endeavoured to sneer it away as a typical preoccupation of the British upper middle class. This, while illustrating intellectual bias, is irrelevant to Darwinian evolution.[9]

My own work in the wild on characters which a competent evolutionist has actually said in print could not matter to their possessors, showed instead that they could be, and not infrequently were, matters of life and death.[10] I was fortunate in having the opportunity to do such long-term research; my younger colleagues are less lucky because such work is often very chancy, and useless as a subject for grant applications since one cannot promise results in one or two years. It is extremely difficult to guess what is

[7] J. A. Endler, *Natural Selection in the Wild* (Princeton, NJ, 1986).

[8] R. C. Lewontin, 'Testing the Theory of Natural Selection' (Review of E. R. Creed (ed.), *Ecological Genetics and Evolution*), *Nature*, 236 (1972), 181–2.

[9] A. J. Cain, 'Introduction to General Discussion', in J. Maynard Smith and R. Holliday (eds.), *The Evolution of Adaptation by Natural Selection* (London, 1979), pp. 599–604.

[10] Cain, 'The Efficacy of Natural Selection in Wild Populations', in C. Goulden (ed.), *The Changing Scenes in Natural Sciences, 1776–1976, Academy of Natural Sciences Philadelphia Special Publication*, 12 (1977), 111–33; id., 'Ecology and Ecogenetics of Terrestrial Molluscan Populations', in W. D. Russell-Hunter (ed.), *The Mollusca*, vol. vi (London, 1983), pp. 597–647.

important for fitness in the home life of so strange an organism as a snail, and to frame one's investigations accordingly; it is far easier, especially if one is under pressure to publish in order to keep one's job, to do some superficial surveys in the wild, to announce that one can find no obvious association between particular genetic characters and any environmental variables, and publish one's conclusion that all the genetic variation is due to chance. As Diderot remarked in 1753, '. . . par malheur, il est plus facile et plus court de se consulter soi que la nature.'[11] *Some* variation in wild organisms is surely due to chance; but where any serious investigation has been done, considerable natural selection has been found, as Endler's book shows. And of course if one decides on insufficient evidence that a particular example of variation is due to mere chance, that is the finest possible way of blocking further work—chance phenomena are merely chance phenomena, and nothing further can be done (except to publish).

Many molecular biologists, and mathematicians, who have no practical knowledge of animals in the wild, have made great play with ideas of non-adaptive evolution. Every variation in the molecular basis of heredity for which there is no obvious function is immediately proclaimed as an example of non-Darwinian evolution. In some examples, as when different codes may produce *exactly* the same protein, there may indeed be a considerable amount of noise in the genetic system. It may be that this can accumulate, at least over long periods, in an approximately linear way with time, in which case it could be of great use as an independent biological clock. No one who has had experience of living animals in the wild over many generations can be happy with such interpretations, which at present are as facile as the hasty conclusions discussed above. It takes a great deal of hard work to get at even the basic natural history of characters or organisms in the wild, and far more to produce an interpretation worth considering (and how damning to any grant application would be the expression 'mere natural history!').

Of course, any hint that Darwin got it wrong has instant news value, and is exploited by so-called science correspondents. There can hardly be a worse fate for any serious and difficult scientific

[11] D. Diderot, 'De l'interprétation de la nature' (1753), in P. Vernière (ed.), *Œuvres philosophiques de Diderot* (Paris, 1964).

controversy than to get taken up by the popular media (except of course to be abolished by ignorant politicians in a cultural revolution).

Dawkins, in *The Blind Watchmaker*,[12] expresses surprise at the widespread opposition to evolution. 'For reasons that are not entirely clear to me, Darwinism seems more in need of advocacy than similarly established truths in other branches of science.' One reason is, I think, obvious: it strikes directly at the personal importance of everyone and (in conjunction with other theories) may even take away the hope of compensation for present injustice in a future life. Natural selection is hated even more than evolution. The classic statement is in the preface to George Bernard Shaw's *Back to Methuselah*.[13] Evolution he could take if it was spiritual, 'a mystical process, which can be apprehended only by a trained, apt, and comprehensive thinker' (note the self-flattery). But natural selection never!—'a ghastly and damnable reduction of beauty and intelligence, of strength and purpose, of honor and aspiration, to such casually picturesque changes as an avalanche may make in a mountain landscape, or a railway accident in a human figure . . . the universal struggle for hogwash.' Excellent writing; emotion, not reasoning.

Human Evolution

The philosopher S. R. L. Clark has rejected the 'standard' account of human evolution for two reasons.[14] First,

the existence of consciousness is incomprehensible if we are merely complex, self-replicating kinetic systems selected for their inclusive genetic fitness over some four thousand million years. Consciousness, the subjectivity of being, can play no part in the evolutionary story.

Second,

even if neo-Darwinian evolution had thrown up conscious beings, it could not be expected to produce creatures with a capacity for understanding the workings of the universe. Cleverness, even verbal and mathematical intelligence, are devices which, under some circumstances,

[12] R. Dawkins, *The Blind Watchmaker* (Harlow, 1986).
[13] Shaw; *Back to Methuselah: A Metabiological Pentateuch* (London, 1929).
[14] S. R. L. Clark, *From Athens to Jerusalem: The Love of Wisdom and the Love of God* (Oxford, 1984).

might have a genetic advantage: who could have foreseen that such practical skills would have so vast an application, so powerful a thrust?

But it will be noted that neither of these objections demonstrates a contradiction in terms or things; incomprehension by a particular person may be of historic interest, but is no argument, the more so because it is by no means certain that the human mind does understand 'the workings of the universe' *tout court*. Certainly it is by far the finest example of preadaptation in the whole animal kingdom. If you can count up to five on one hand, you can count up to ten with both, you can bend down fingers for subtraction, you can show repeatedly combinations of extended fingers for multiplication, and most of the possibilities of number theory can be developed from such simple operations. Alfred Russel Wallace insisted that musical and mathematical ability could not be evolved by savages (and became a spiritualist in consequence).[15] T. H. Huxley's rejoinder to him[16] still seems to be valid—indeed it largely coincides with Leakey's and Lewin's assessment[17] of the evolution of human mental and cultural abilities. Moreover, if the human mind has been evolved, one would expect it to be as *ad hoc* as any other evolved device, and incapable of conceiving what is not either of direct use to it or a logical consequence of what is— such as the nature of consciousness.

The fossil record of human and hominoid evolution, although of course fragmentary, is far better than for such other mammalian orders as pangolins, aardvarks, and monotremes. A useful review is that by Leakey and Lewin; new discoveries are made almost monthly but they do not affect the broad picture. From comparative anatomy, Man belongs with the great apes and the extinct australopithecines in a subdivision of the order Primates, itself only one of the thirty-odd orders of the true mammals (Class Eutheria) which are only one class of the vertebrates. Mammals, birds, reptiles, fish are so different that non-specialists cannot believe there to be less difference between them than between starfish, and jellyfish. The bulk of the multitudinous diversity of

[15] A. R. Wallace, 'The Limits of Natural Selection as Applied to Man', *Contributions to the Theory of Natural Selection* (London, 1870), chap. 10.

[16] T. H. Huxley, 'Mr. Darwin's Critics', *Critiques and Addresses* (London, 1871; repr. 1883).

[17] R. Leakey and R. Lewin, *People of the Lake* ... (Harmondsworth, 1981).

animals lies within those groups usually dismissed as mere invertebrates. There never was an evolutionary tree; it was always a bush, clump, or thicket, with Man as one small twig. Here Julian Huxley's terminology is useful.[18] Anatomically Man is only one of numerous mammalian *clades* (evolutionary lines). But in his development of mind and cultural evolution he surpasses by far every other known animal; and Huxley placed him as the only occupant of a new *grade* in evolution, the Psychozoa. Cultural evolution is not confined to Man but he has developed it so far as to require the formal expression of this difference. It is very difficult to find any absolute difference between Man and other animals—perhaps his possession of a financial as well as a physiological metabolism? Dr Dunbar suggests to me that 'formal religion is the only one. Animals take the world as they find it; we have to account for it.'

Probable Consequences

If, then, Man has evolved like other living things, what character-istics can we, from our general knowledge of evolution, expect him to show as an animal in general, and as one sort of animal in particular? Like every living thing he will have an intense drive to ensure his individual survival and to reproduce; without them, his lineage could never have survived for three thousand million years. And like every other species, he will have been selected to be *comfortable* in his environment. An individual always dissatisfied and attempting to escape is unlikely to persist and reproduce in the wild.

One of the most important sources of comfort is a firm belief in one's personal importance, and the ability to produce rationaliza-tions of conduct, wish-fulfilment fantasies, and the like compensa-tions when things do not go as one would like. After all, very few males (for example) can dominate a group all the time and have everything their own way.

Erasmus Darwin's 'three great objects of desire, which have changed the forms of many animals by their exertions to gratify

[18] J. S. Huxley, *Clades and Grades*, in A. J. Cain (ed.), *Function and Taxonomic Importance* (London, 1959), pp. 21–2.

them . . . lust, hunger, and security', cover a great deal of necessary action for survival. Charles Darwin rightly remarked in 1876[19] that

pain or suffering of any kind, if long continued, causes depression and lessens the power of action, yet is well adapted to make a creature guard itself against any great or sudden evil. Pleasurable sensations, on the other hand, may be long continued without any depressing effect; on the contrary, they stimulate the whole system to increased action. Hence it has come to pass that most or all sentient beings have been developed in such a manner, through natural selection, that pleasurable sensations serve as their habitual guides.

Just what actions are pleasurable depend in large part on what sort of life an animal leads—what sort of ecological job it does. The evidence indicates pretty clearly that Man evolved in a savannah environment, in groups larger than the immediate family, and with a considerable overlap of generations, the young being dependent on their elders for some time. There is much physical and mental change as the young grow up, but nothing like the metamorphoses of moths, starfish, or tunicates. The wide vistas, great seasonal changes, and diversity of scenery, vegetation, and foodstuffs that are characteristic of a watered savannah are our natural setting. It is noticeable that the British aristocracy at the height of its power, wealth, and culture in the eighteenth century made savannah landscapes of its estates—open vistas with clumps and belts of woodland, streams, or lakes, beds of highly seasonal flowers—as did every other such social class in a seasonal environment that I know of. Not one embedded his mansion in dense forest or in a swamp or lake (a moat for protection hardly counts against my proposition). Not to have such variety and change produces boredom, an acute form of mental pain—but it can hardly affect the clothes moth larva or wood-boring beetle grub, living in an environment of near-total monotony.

To use a diversified and changing environment for personal survival requires some expenditure of energy daily, and there are plenty of other calls for activity, e.g. in attracting a mate. Pascal is reputed to have said that most of the troubles of the world arise

[19] C. R. Darwin, *The Autobiography of Charles Darwin*, (1876), ed. N. Barlow (London, 1958), p. 11.

because men are not content to sit quietly in a room. Perfectly true, but useless as a recipe for inaction.

In a group, there is a constant conflict between immediate advantage (grabbing what one wants) and longer-term advantage; keeping friends to help one against aggression; persuading a member of the opposite sex to mate; getting warning from others of the approach of predators, for example. In Man, at least one sex must devote time and energy and food to the offspring, and educate them. Offspring must learn all they can while still protected, and exercise and play with others, since physical development is partly a matter of exercise. In such a social environment the capacity to learn is essential, and surely must have a considerable genetic component. Similarly some water plants, faced with unpredictable variations in water level, have a repertoire of leaf forms—divided leaves, having little resistance to water-flow, for the submerged plant, entire leaves with a bigger surface area for photosynthesis when the plant is exposed to the air. Different items of a repertoire are evoked by different environmental stimuli, but the capacity for the repertoire is inherited. What is to be known by an individual primate in a group includes not only the unique environmental situation of that group (situation of safe watering-places, refuges from predators, feeding-grounds for different seasons, etc.) but the moment-to-moment changes of its relationships with the other members of the group. As Dunbar has illustrated forcibly, one individual's behaviour cannot be seen in isolation from that of other members of its group.[20] Almost none of what this requires can be programmed or elicited by all-or-none signals—very unlike the situation in social insects in much of their social life (but not in foraging for food). Two major behavioural features stemming from group life are an acute awareness of one's relative social position (and of that of potential rivals and mates) and an awareness (apparently) of kinship.

All these activities, or rather the capacities for them, will have been improved by natural selection, but to what extent is extremely difficult to discern in Man, since variation induced by

[20] R. I. M. Dunbar, 'The Evolutionary Implications of Social Behaviour: The Problem of Optimal Strategy Sets', in H. Plotkin (ed.), *The Role of Behaviour in Evolution* (London, 1988), pp. 165–88.

learning is so marked a feature of mankind. Moreover, with spreading fundamentalist and other reaction (in more than one religion) against rationality and science, and with the irrational prejudices of scientists (who are, after all, only human) the subject-matter itself may revolt and refuse to allow the intending investigator to work. Yet one of the most urgent problems in human behaviour is to determine how much is genetic and how much is effectively imprinted during early life, and thereafter merely defended, not modified by learning.

With the great reduction (but not abolition) of natural selection in highly civilized societies, plus the psychological strains of living in vast communities and a highly artificial environment, all the normal activities of the individual can go wrong, and we find non-reproducers, sadists, masochists, shoe-fetichists, suicides, and so on; with modern communications, they can get in touch with other like-minded individuals and give each other mutual support. Some of these evolutionary eccentrics may make a contribution to cultural evolution by forcing reconsideration of traditional values in a changing world, or setting new ideals of conduct. Others seem useless from any point of view.

It has long been a commonplace of philosophical and theological criticism of evolution that it has no place for values. Many scientists have taken the same stand. The psychologist William McDougall asserted that 'it is the nature of man to recognize the true, the good, and the beautiful, to esteem highly all such things, to aspire towards them, to strive to preserve, augment, and create truth, goodness and beauty'[21] and that these supreme values 'are independent of Science in the sense that no conceivable discoveries made by scientific methods can refute or shake them.' But beauty, at least, is certainly relative to our comfort in our own environment, and our appreciation of the health, vigour, and physical characteristics of a potential mate. To beetles, other beetles must appear beautiful. I think our appreciation of goodness and truth can also be shown to be largely relative to our own particular evolutionary history, although truth in the sense of correspondence to actual experience must always have been a necessity for all organisms; what an individual chooses to believe about his

[21] W. McDougall, 'Religion and the Sciences of Life', *South Atlantic Quarterly*, 31 (1932), 15–30.

ultimate importance, the meaning of the world, and suchlike matters need have no relation to experience at all.

Cultural Evolution

Cultural inheritance and evolution are being examined with much more care and accuracy than ever before, but the study is very recent. In an excellent attempt to provide suitable mathematical theories for modelling cultural evolution, Cavalli-Sforza and Feldman have transposed and modified the mathematics of population genetics for use in this study.[22] They rightly bewail the lack of data, but analyse several studies of outstanding interest, e.g. of assortative marriage by religion and cultural inheritance of religious attitudes.

When we compare cultural with Darwinian evolution, we find several obvious contrasts. (i) Cultural inheritance can be from all directions, not just from parents. (ii) The old and even the post-reproductive (the *expended* phenotypes) may be highly important in the education of later generations—from my own experience, this seems, after early childhood, to be less important in Western culture now than in the past and in other cultures. (iii) Potentially, cultural evolution is extremely fast, especially with modern devices for mass communication, though it is one thing to receive a message, another to understand it, and a third to accept and act upon it. (iv) Both potentially and actually, cultural change may be orthogenetic or Lamarckian—a need is felt and there is a deliberate attempt to produce innovations to satisfy it and to improve them subsequently. No doubt much cultural change is comparable with much genetic mutation, mere noise, but much is not and has no parallel in Darwinian evolution. And (v) we see, in some cases, as a result of cultural evolution, what appears to be true group selection, e.g. the extinction of the Tasmanians by invaders with a very different and aggressively efficient culture.

These potentialities are realized to amazingly various degrees. Technological changes which give immediate profit are comparatively eagerly seized on; so are fashions in clothing or ideas,

[22] L. L. Cavalli-Sforza and M. W. Feldman, *Cultural Transmission and Evolution: A Quantitative Approach* (Princeton, NJ, 1981).

which give an immediate boost to the ego, but these are much more culturally limited. Religious beliefs, which affect one's self-importance and solidarity with one's social group so profoundly, are exceedingly slow to change. And indeed, for most of the human race, cultural inheritance is conservative, not innovative. Theophrastus's character of the superstitious man (*c*.315 BC) is still only too topical over most of the world.[23] Signals, symbols, and information may be bounced off satellites in nanoseconds, but the recipients have not changed like the technology. No very extensive experience is needed to find that emotions come first in determining most of human conduct, that reasoning, except in immediate practical concerns, is mostly used to rationalize, and that short-term interests take precedence of long-term advantages. In few matters is this clearer than in the general failure of campaigns to advocate birth control, even in countries where a large family is not an immediate economic advantage.

The contrast between the potential rapidity of cultural (perhaps only technological) evolution and the slowness of genetic evolution is certainly striking. But what is of far more immediate importance is that between cultural evolution and the rate of human reproduction. Without any nuclear holocausts, it is more than likely that at least the terrestrial parts of the planet will be ecologically and economically ruined in half a century simply because cultural evolution is too slow. Although Man is unique in his degree of employment of symbolic communication, the symbols that arouse the greatest emotions, and therefore that lead to action including the rejection of other symbols, are not appropriate to provoke a sudden voluntary check in population growth rate. They work only with those, few and mostly politically impotent, who can take a long view, make the consequence vivid to themselves, and act accordingly. Intellectuals are always prone to believe that the rest of the world are as open to reason as they (but not their colleagues) believe themselves to be. But, as Lawton has said, no real politically viable alternative has been presented to the multiplying myriads of the world, which would protect the deteriorating environment.[24] It is no good standing in line, singing 'Woodman, spare that tree!' to a shifting

[23] *Theophrastus: Plays and Fragments*, trans. P. Vellacott (Harmondsworth, 1973).
[24] J. H. Lawton, 'Newts and Nuclear Winters', *Nature*, 321 (1986), 390.

cultivator with a wife and children to feed, who is about to demolish the last patch of rain-forest in his neighbourhood.

The True Meaning?

It would appear, then, that the human race has been evolved in circumstances that have given it, for the most part, an amazing flexibility of behaviour and power of communication in immediate practical matters for immediate material advantage, but with far less flexibility in matters concerning personal status, production of families, and support of one's immediate group than is desirable (indeed essential) to prevent even short-term deterioration of the world's environments. What with an undefined and perhaps indefinable genetic burden mediating but also limiting our capacities for acting for long-term advantage, and the slowness of so much of cultural evolution, probably mediated by something like imprinting in early childhood, the prospect is bleak indeed. The only ray of hope is that appropriate symbols will be found to alter the behaviour of a sufficiently large proportion of the human race to avert disaster. Lawton remarks: 'The price of failure doesn't bear thinking about.' Thinking about it may be the only thing that will convince people in general, but as he also says: 'How can we possibly understand the machinery of nature on miniscule [*sic*] budgets that are a direct result of government indifference to environmental science?' And he is right.

The Universities

Not much scientific research can now be carried out successfully by independent amateurs, even if very wealthy. Almost the only institutions in the world that *can* do disinterested research into what we need to know for long-term survival are the universities; but, because of the need of money for research, far too many of them are in the hands of hard-line governments who require rigorous toeing of a party line and the production of subservient technicians to understand, handle, and design complex equipment for immediate gain—in short, brainwashing, not that form of education which teaches people to think for themselves. Intellectuals no doubt are individually biased; nevertheless by their intellectual conflicts with one another and, in science, by appeal to actual experiment, they can progress towards a more objective appreciation of the world and Man. What they have to realize is that they are more probably among the evolutionary eccentrics, having (some of them at least) remained children in their curiosity and exploration of the world, than representative of mankind in general. Periods in history when discussion has been genuinely free are very few, very short, and often followed by repression or downright persecution, as Dodds has shown for that apparently most rational of all groups the thinkers of Periclean and post-Periclean Athens.[25]

It is more than possible that in Britain we are coming to the end of one such golden period; Dorothy Nelkin's book on the conflict between fundamentalism and science in the USA documents the immense strides made by irrationalism in that country of glorious freedom.[26] Certainly the culturally retarded politicians that can see in a Vice-Chancellor only a works manager and in a university only a source of cheap technological information; who in the name of efficiency are removing all safeguards to intellectual freedom and concentrating power in the hands of a few to ensure 'efficient' running; and whose only criterion of value in appraising proposed research is the amount of money successfully acquired and spent on a piece of work; are very unlikely to respond to any long-term plans. There will be enough quislings inside the universities to abet them.

Prospects at Different Times

In one respect William McDougall was certainly right. We are here, with our friends, families, rivals, competitors, all of us responding to the good, the true, and the beautiful, the inconvenient, the aggressive, the intriguing even if we interpret those terms in different ways at different times, and don't always live up to

---

[25] E. R. Dodds, *The Greeks and the Irrational* (Berkeley, Calif. and London, 1951).
[26] Dorothy Nelkin, *Science Textbook Controversies and the Politics of Equal Time* (Cambridge, Mass., 1977).

what we say. We are too embedded in the present for knowledge of origins to have much relevance to our daily lives, though in a few educated people it may persuade them to extend more freely to others the tolerance they (naturally) demand for themselves. By preadaptation, the human mind is capable of attempting to foresee the near future (though a very small number actually do). Like all other human activities this can go wrong (Tarot packs, tea leaves, palmistry) and at best there is room for doubt—I may be quite wrong about the slowness of crucial elements in cultural evolution. One sympathizes with the Irishman who said he would rather prophesy after the event. Really long-term events are beyond us. I am told by a professor of theoretical physics that even the entropic death of the universe is not quite certain; a calculation in the late nineteenth century showed that there is a finite probability that order will slowly build up again out of chaos, but the probability is so minute that even cosmologists ignore it. It is only the more immediate prospects that stir us to thought, let alone action—just what one should expect of an organism evolved *ad hoc*.

2

Evolution in Sociology

W. G. RUNCIMAN

In 1937, when Talcott Parsons opened his book *The Structure of Social Action* by quoting from Crane Brinton the rhetorical question, 'who now reads Spencer?', he can hardly have expected that he would live not only to see Spencer's reputation as a sociologist rise triumphant from the ashes but to hear the question asked, 'who now reads Parsons?'. Such reversals of fortune are common enough in the history of ideas. But why did Spencer's sociology go so soon and so completely out of fashion? Was it because it was displaced by Marxism and reactions to Marxism? Was it because of the repudiation by the rising generation of anthropologists of conjectural history and the so-called comparative method? Was it because of a philosophical reaction against the mechanistic presuppositions which underlay his whole conceptual system? Was it because of a growing distaste for his extreme individualism in matters of social policy? Or was it a mutually over-determining combination of them all?

The answer, as to all such questions, lies partly in the strength of the criticisms brought directly to bear against Spencer's own arguments and partly in changes in the social and political climate within which these criticisms found an increasingly receptive hearing. With hindsight, the story can be told in such a way as to compound the ironies to the point of tragicomedy. Spencer did not lack critics in his lifetime—to whom his alleged response was to finger his pulse and say, 'I must talk no more' when being worsted in argument.[1] Yet in his last years he sank into a profound despair both about the future of civilization and about the fate of his own work. Which, one wonders, would have surprised him more: that only two generations later he should come to be hailed as having 'not only introduced the concepts of what is now called "structural-

---

[1] The testimony is Galton's citing 'wicked friends': J. D. Y. Peel, *Herbert Spencer: The Evolution of a Sociologist* (London, 1971), p. 31.

functionalism", but also laid foundations for a systematic analysis
of social phenomena',[2] or that Karl Marx, opposite whom he is
buried in Highgate Cemetery, should have come to be hailed
likewise as the author of a doctrine which 'is far from the only
structural-functionalist theory of society, though it has good
claims to be the first of them'?[3] By this time, what is more, the
notion of 'function', which in Malinowski's hands had become the
tool of a social theory not only static but circular, had recovered
the dynamic implications which both Spencer and Marx had
assigned to it, and American theorists of 'development' and
'modernization' were cheerfully usurping without acknowledge-
ment the evolutionary tradition in sociology which Spencer had
largely founded. Even Parsons is to be found in 1961 editing *The
Study of Sociology* for the University of Michigan Press and in
1966 publishing a mercifully short book of 115 pages under the
title *Societies: Evolutionary and Comparative Perspectives*. And in
1968 Spencer's influence on modern British sociology was
acknowledged by Philip Abrams as 'decisive', if only because it
had been built as a defence against him. As Abrams says, 'Spencer,
the sociologist who first made a systematic analysis of the
unintended consequences of social action, was himself a victim of
the process he studied.'[4]

Now selective quotations and tendentious anecdotes are not, I
am well aware, the stuff of which serious intellectual history—
even potted history—can be made. Even in a Spencer lecture it is
only right to make clear that Spencer was not the only person to
introduce the notion of evolution into sociological theory and that
there are influences other than his which have kept it there. But the
ironies are genuine, and there are genuine lessons to be drawn
from them. They arise partly, as I have already implied, because of
changes in political rather than intellectual fashion: Spencer's
astonishing reception in the United States, in particular, was so
clearly linked to the condition of post-bellum America with, as
Richard Hofstadter put it, 'its rapid expansion, its exploitative
methods, its desperate competition, and its peremptory rejection of

---

[2] Stanislav Andreski, 'Introduction' to his edition of Spencer's *Principles of Sociology*
(London, 1969), p. xiii.
[3] Eric J. Hobsbawm, 'Karl Marx's Contribution to Historiography', *Diogenes*, 64
(1968), p. 46.
[4] Philip Abrams, *The Origins of British Sociology, 1834–1914* (Chicago, 1968), p. 67.

failure',[5] that his ideas were bound to be exaggeratedly admired then and exaggeratedly denounced later, when the world failed to conform to his individualistic and peaceable predictions. But to the extent that it is possible to isolate a single intellectual problem which bedevilled the argument over social evolution throughout, it lies in the connection between evolution and progress. All the protagonists are agreed that evolution is something more than mere qualitative change. Indeed, it is two things more. It is change in some sense or other for the better and it is change in the direction of a goal. From neither implication did Spencer's critics escape any more than his successors. They neither wished nor attempted (nor would have been able if they had) to extricate themselves from the toils of a concept whose attraction derived from the very presuppositions which vitiated its usefulness for a sociology which could lay definitive claim to the style and title of social science. In this lecture, accordingly, I propose to consider each implication in turn before going on to suggest to you that the task which Spencer set himself as an evolutionary sociologist is, all the same, capable of fulfilment still.

The idea of progress is such a familiar item in any inventory of the furnishings of the mid-Victorian mind that it may seem that in focusing my attention on its connection with the idea of social evolution I am doing no more than highlighting something which has been amply illuminated already. But what I have read, or reread, in the course of preparing this lecture has convinced me that it is one of those aspects of the history of social thought which cannot be emphasized too much. It is a product of that world of railways, geology, religious doubt, patriotic enthusiasm, the Crystal Palace, domesticity, hypocrisy, sentimentality, reformism, complacency, and the 131 Stanzas of *In Memoriam* which is at once so familiar and so alien. How *could* Tennyson manage so unselfconsciously to combine (as G. M. Young put it) an unwillingness to quit with an incapacity to follow any chain of reasoning likely to lead to an unpleasant conclusion?[6] How *could* he so unselfconsciously confuse (as T. S. Eliot put it) the hope of immortality with the hope of the gradual and steady improvement

[5] Richard Hofstadter, *Social Darwinism in American Thought* (2nd edn., Boston, 1955), p. 44.

[6] G. M. Young, *Victorian England* (2nd edn., London, 1953), p. 75.

of this world?[7] But he did. Somehow, for all that science has taught us and religion has abandoned to science, mankind is going to come through all right in the end. And it is not just the sociologists—Spencer or Marx or the devotees of the Comparative Method—who clung so hard to a belief in progress. The belief extends, as Messrs Collini, Winch, and Burrow have recently reminded us, all the way from the shrewd, sensible, worldly, ironic Bagehot to Alfred Marshall, Keynes's 'first professional economist', whose *Industry and Trade*, published as late as 1919, can, not unfairly, be read as a Whig version of economic history of which the 'business point of view'[8] is the theme and the sturdy English race the evolutionary hero. It is all (to us) so earnest, yet so muddled; so concerned, yet so sanguine; so intelligent, yet so blinkered; so knowledgeable, yet so gratuitously moralistic. And it has bequeathed an intellectual legacy which, for better and worse, has not been dissipated yet.

Lecturing as I am in the home of lost causes and forsaken beliefs, I have no need to remind you that Matthew Arnold (whose views about perfection in style Spencer emphatically did not share) was the Professor of Poetry who said in 1865 that 'it is a result of no little culture to attain to a clear perception that science and religion are two wholly different things.' Times, to be sure, have changed a little since then. It is not Bishop Colenso's commentary on St Paul's *Epistle to the Romans* which elicits the attention of the young barbarians when not at play, but Professor Cohen's on Marx's *Capital*. But if for religion we read ideology, times have not, perhaps, changed so much. It is not that Spencer himself, any more than his critics or successors, ever claimed that judgements of value and statements of fact are not two different things. Even Hobhouse, in his opening editorial article for *The Sociological Review* in 1908, is categorical that 'sociological thinking must start with a clear cut distinction between the "is" and the "ought", between the facts of social life and the conditions on which society actually rests and the ideal to which society should conform.' But then in the very next breath Hobhouse goes on to say that it is

[7] T. S. Eliot, *Selected Essays* (2nd edn., London, 1934), p. 335.

[8] Stefan Collini, Donald Winch, and John Burrow, *That Noble Science of Politics: A Study in Nineteenth-Century Intellectual History* (Cambridge, 1983), pp. 169, 328. The reference to the 'business point of view' is taken from *Industry and Trade* (p. 163) where Marshall refers to its emergence as a 'chief feature of economic evolution'.

impossible to keep sociology 'in permanent separation from all ethical considerations';[9] and in his own writings, as is well known, he sought to develop the notion of what he called 'orthogenic evolution'—a process, as he saw it, whereby as the social mind develops the prospects for harmony and justice gradually improve. It may still be possible to argue, as one of his recent commentators has done, both that Hobhouse had 'a viable theory of social development of which his pupil Morris Ginsberg offered an able defence in later years' and that his 'subsequent sociological standing might well have been higher if he had cut loose altogether from the terminology of evolution.'[10] But Ginsberg's defence, however able, was not able enough; and Hobhouse never did cut loose from the theory of evolution as progress, any more than did his anti-liberal, Marxist counterparts. For all the claims that have been made, from Engels to Althusser, for the scientific status of Marx's theory of social evolution, it is Marxists themselves who have been most explicit that socialism is a better state of society than capitalism and that the function of Marxism is not only to predict it but also to further it as a goal. Nor, for that matter, is the implication that evolution is for the better any less explicit, even if the word itself is not used, in the writings of the American theorists of 'modernization' whom I have just described as the cheerful usurpers of the Spencerian tradition. Professor S. M. Lipset's often-quoted assertion in 1959 that democracy (meaning, of course, Western liberal democracy) is 'the good society itself in operation'[11] does no more than articulate in the starkest terms what is implicit in the writings of a whole school whose satisfaction with the institutions of their own society in their own time matches that of even the smuggest mid-Victorian Englishman.

The trouble in all this is not that sociologists ought somehow to be debarred from talking at all about progress from worse to better. If, for a given society over a given period, it can accurately be reported that infant mortality has fallen, real wages per head have risen, and literacy has spread, I do not suppose that anyone in this audience will insist on striking out from the publication in which the report is made any word which even hints at the notion

[9] L. T. Hobhouse, 'Sociology, General, Specific and Scientific', *Sociological Review*, 1 (1908), pp. 4–5.

[10] Peter Clarke, *Liberals and Social Democrats* (Cambridge, 1978), p. 148.

[11] S. M. Lipset, *Political Man: The Social Basis of Politics* (New York, 1959), p. 403.

that these changes are at the same time an improvement. But that
is not because we all subscribe to a philosophical theory whereby
good and bad are attributes of states of society ascertainable by
observation. It is because of the contingent connection between
values which we happen to share and the reported social facts.
This, let me emphasize, is not 'value-relevance' in Max Weber's
sense, for Weber's criterion (as I have argued elsewhere)[12] is
tenable only if it overrides the distinction between the sciences of
man and of nature which it was Weber's purpose to sustain. It is
simply a recognition that we do all have values about states of
society which we do not about states of inanimate matter—from
which it follows that the changes which sociologists report and
seek to explain are indeed likely to have implications to which,
according to our particular values, notions of good and bad,
success and failure, or progress and decline will attach. But it does
not follow that we are in any way bound by what sociologists of
one school or another have reported and explained to us to accede
to *their* judgements of value. It is logically possible to agree with
Spencer that societies have evolved from militancy to industrialism
without agreeing that welfare legislation is a mischievous inter-
ference in the elimination of the unfit; it is logically possible to
agree with Marx that capitalism will evolve into socialism without
agreeing that profit made out of the purchase at the market price
of the labour of formally free wage-workers is wicked; and it is
logically possible to agree that the empirical correlation between
'economic development' and 'democracy' as he defines them is
what Professor Lipset says it is without agreeing that democracy
so defined is indeed the good society itself in operation.

But if progress is for the better only if your particular values
happen to make it so, is it not still progress towards a goal? The
problem here is not, strictly speaking, that of teleology. For
Spencer, as for Marx, it is not the properties of the end-state which
explain the evolutionary process by which it comes about. Yet one
is tempted to say that it might as well be.[13] It is true that Spencer

---

[12] *A Critique of Max Weber's Philosophy of Social Science* (Cambridge, 1972), chap. 4.
[13] Thus David Wiltshire, *The Social and Political Thought of Herbert Spencer* (Oxford,
1978), pp. 206–7: 'Spencerian social evolution, then, anticipates a future in which
continuing trends culminate in perfection. This inevitably raises the question of moral
purpose in evolution; it is essentially a teleological view of human development. The stages
of the theory at which science and morality collide are the concepts of the unknowable (the

did not conceive of evolution as caused by the pull, as it were, of the millennium but by the big bang of natural causation which pushes human societies, despite regressions and setbacks, in that direction. It is also true that Hobhouse followed Spencer in seeing social evolution as a special case of the cosmic evolution set off by an unknowable first cause. Indeed, even Marx, who can, perhaps, plausibly be charged with teleological reasoning, still holds the contradictions between forces and social relations of production which, on his theory, bring about the evolution from one mode of production to the next to be pushed from behind rather than pulled from in front. Yet for all that they cannot plausibly be labelled unilinear evolutionists, both Spencer and Marx are equally explicitly aware of an end-state to which they see the processes of social evolution as tending—an end-state, moreover, in which according to both of them, government by force will have given way to administration by co-operation. Looked at from the far end, the difference between Spencer's and Marx's evolutionary sociology is far less than when looked at in the context of their diametrically opposed evaluations of individualism and collectiv-ism. For both, a harmonious and superabundantly productive division of labour is the ultimate goal; and for both, the evolution of human societies towards it is no more a matter of chance than it is of the working-out of the will of an all-wise God.

Here, therefore, the trouble is simply that the right answer *is* chance—or rather, that the opposition between chance and determinism is not what nineteenth-century social evolutionists of all schools presupposed it to be. To say that the process is random is not to say that it is uncaused, any more in the evolution of societies than of species. It is only to say that it is caused independently from and unpredictably with reference to the outcome which calls to be explained. If, to speak thermo-dynamically, social evolution has thus far run uphill, it is not because there is an inexorable series of progressions from chaos to order or from lower to higher output of energy per head. It is because the potential for variation in social structure and function is present in the biological inheritance of the human species, and

point at which morality enters the system) and of adaptation, in which an empirically-defined process is with great difficulty reconciled to the assertion of evolutionary perfectibility.'

the interactions of members of different societies with their natural environment and each other has happened to produce the particular outcomes which we observe, in which some organizational forms dominate some others. Because resources are finite, the process is by definition competitive. But it is not a knock-out competition, and the pattern of domination (or co-operation) between societies of one organizational form and another as it is at any one time is neither inevitable nor irreversible. We can continue to talk about evolution because the outcome which the process has generated is, as in the evolution of species, one in which new varieties continue to appear and to displace, preserve, or modify the old ones. But the presupposition which has to be discarded is that, as it is put and endorsed by J. D. Y. Peel in his useful and perceptive study of Spencer and his sociology, 'Social evolution, in any precise sense, is only appropriate when we really are justified in asserting that a particular outcome is necessary.'[14] That is simply not so; and the failure of numerous Spencerians and Marxians alike to acknowledge that it isn't has been as much of an impediment to the advance of sociological theory as has their failure to acknowledge that the connection between sociology and ethics is a purely contingent one.

So where does this leave us? If sociology, although both evolutionary and scientific, is neither of these things in a way that either Spencer or Marx would recognize, what is it? Here again, times have not changed so much as one might have hoped. In Hobhouse's editorial article of 1908, from which I have already quoted, he complained that 'Not only are there still many who deny the bare existence of Sociology, but, what is more serious, among Sociologists there are still many deep divergencies of view as to the nature and province of the enquiries which they professedly pursue in common.'[15] The question 'what is sociology?' is not, perhaps, asked as frequently or as aggressively in Oxford or Cambridge as it was thirty years ago. But its numerous practitioners are hardly less divided about the answer than they were in 1908; Sir Isaiah Berlin would, I suspect, still maintain as he did in 1953 that 'it is a commonplace to say that sociology still awaits its Newton, but even this seems much too audacious a claim; it has yet to find its Euclid and its Archimedes, before it can begin to

[14] Op. cit., 257.   [15] Op. cit., 1.

dream of a Copernicus';[16] and the unkind rhetorical question to which sociologists are apt nowadays to be subjected is, 'what is there in sociology which is neither second-rate history nor second-rate philosophy?' (to which the even unkinder answer is 'second-rate journalism'). Yet it would be merely flippant if not actually perverse to suggest that the subject, under any definition, is in as sorry a state as that. As I said at the beginning, there are genuine lessons to be learned from the ironies of hindsight, and many of them have been. The subject-matter of sociology is and always has been what Spencer said in 1873: '. . . the growth, development, structure and functions of the social aggregate as brought about by the mutual actions of individuals';[17] and its task is to analyse these not quite as either Spencer or Marx did, but in the light of both their insights and their mistakes.

As it happens, I have addressed myself to very much this assignment in a Radcliffe-Brown lecture to the British Academy delivered earlier this year under a title which presumptuously but deliberately echoes that of Darwin's joint presentation with Wallace to the Linnaean Society in 1858. It would, despite Spencer's acknowledged influence on Radcliffe-Brown,[18] be clearly inappropriate for me to use this lecture to argue my own case. But if Spencer and Marx were both right in seeing that sociological theory, if it is to establish itself as something more than an amalgam of second-rate history and second-rate philosophy, has to do for the study of societies what Darwin did for the study of species, then it ought by now to be possible in the light of their insights and mistakes to say how it is that societies do evolve—that is, to specify the process by which they change from one distinguishable type or mode to another, the units selected by that process, the functions which the units perform and the direction which the process has taken (or perhaps I should say: has *happened* to take) thus far. Briefly, my suggested answer is that the process is one of competitive selection whereby certain roles and institutions come to replace or supersede others; that the units of selection are not roles or institutions but the practices of which classes, status-groups, orders, factions, sects, communities, age-

---

[16] 'Historical Inevitability', reprinted in *Four Essays on Liberty* (Oxford, 1969), p. 112.
[17] *The Study of Sociology* (Ann Arbor, Mich., 1961), p. 47.
[18] Albeit largely indirectly, through Durkheim: Peel, op. cit., 238.

sets, and so forth are the carriers; that their function lies in
maintaining or augmenting the power which attaches to the roles
and thereby institutions which they constitute and thus in
preserving or changing the mode of the distribution of power in
societies (or 'social aggregates' or 'social formations') taken as a
whole; and that the direction which evolution has thus far taken
has in consequence been one of both increasing and diminishing
variation—increasing as mutant or recombinant practices create
new roles and institutions, and decreasing as the competitive
advantages which they confer on their carriers compel pre-existing
ones to adapt to them.

That, or something recognizably like it, is, I confidently believe,
the outline of the answer to the problem of sociological evolution
as addressed not only by Spencer and Marx but by Comte and
Saint-Simon and before them the thinkers of the Scottish Enlighten-
ment. But even if I am wrong about that, the argument thus far
points clearly to a related conclusion which again vindicates
Spencer's aims although not his particular way of pursuing them.
Any evolutionary theory necessarily implies a taxonomy in which
societies (or 'social aggregates' or 'social formations') are distinguished from one another not only by synchronic differences of
structure and culture but also by their place in a presumptive
diachronic sequence: although the sequence may not be unilinear,
it is no more the case that any type of society can evolve directly
into any other than that any species can. Now Spencer did, of
course, address himself to taxonomy: that was precisely his
purpose in commissioning the so-called *Descriptive Sociology, or
Groups of Sociological Facts*, a set of whose enormous volumes I
found gathering dust in a safe in the London Library. But the
trouble this time was that Spencer failed to restrict himself to the
identification and comparison of those structural and cultural
characteristics which are the materials for the construction of an
evolutionary theory testable against its rivals. He ended up doing
what Sir Edmund Leach was subsequently to disparage (with
Radcliffe-Brown in mind) as 'butterfly-collecting'.[19] Or as John
Burrow has put it, it was 'in pursuit of the remoter ramifications of
social evolution that Spencer's sociology went stamping out into
the forests of prehistory and there, among the all too numerous

[19] Edmund Leach, *Rethinking Anthropology* (London, 1959), p. 2.

variables, died.'[20] Whatever allowances are made, it is truly impossible to conceive how the study of social evolution can be advanced by such categories as 'excessively filthy', 'watchmen on every public building', or 'smoke tobacco till senseless', or by such reported facts as that the Romans pierced their ears for ear-rings from earliest times, that gypsy straw hats were worn in England in 1745–6, that in China imprisonment for using enticing words was abolished in AD 86, that the equanimity of the Eastern Bantu is not disturbed by thoughts of the future, that the Nilo-Hamitic peoples admire white teeth and ample buttocks, or that in some islands the Melanesians are honest to one another but in others the very opposite. But there is no need for a workable theory of social evolution to concern itself with this sort of desperate quest for a *sociologie totale*. It is perfectly feasible for those whose subject-matter is the growth, development, structure, and functions of the social aggregate to steer a middle course between the extremes of, on the one hand, totting up an inventory of any and all observable differences between one society and the next without regard for theoretical significance and, on the other, attaching to societies high-sounding a priori labels in the belief that the workings of their component institutions are thereby automatically explained. If Spencer had extended his notion of the selective evolution of interdependent institutions exerting 'mutual force', as he put it, on one another to an analysis of the concept of power as such, he might both have emancipated his evolutionism from its bondage to distinctions between 'low' types and 'high' and have put a much reduced 'cyclopaedia' of social facts to seriously constructive use. Indeed, Marx's evolutionary taxonomy of modes of production, although it goes too far in the other direction of schematic over-simplification, is all the same a better starting-point—as its continuing influence testifies. It is a common criticism of Spencer that he failed to recognize the importance of the struggle between classes, and although he can be rescued to some degree by selective quotation,[21] the charge must still be allowed to stand. But then it

---

[20] J. W. Burrow, *Evolution and Society: A Study in Victorian Social Theory* (Cambridge, 1966), p. 207.

[21] e.g. *The Study of Sociology*, p. 220: 'The egoism of individuals leads to an egoism of the classes they form; and besides the separate efforts, generates a joint effort to get an undue share of the aggregate proceeds of social activity. The aggressive tendency of each class, thus produced, has to be balanced by like aggressive tendencies of other classes. Large

is an equally common criticism of Marx that he emphasized the struggle between classes (as he defines them) to the exclusion of all else, including the military rather than economic struggles about which Spencer had some perceptive things to say in *The Principles of Sociology* even if he, like Marx, underestimated their importance in his own lifetime. Had they both recognized that power in societies is of three fundamental, interdependent, but mutually irreducible kinds—economic, ideological, and coercive—then they might have arrived at a principle of taxonomy whereby societies are categorized not merely according to their modes of production but their modes of persuasion and coercion too; and they might have succeeded after all in steering the middle course between indiscriminate trait-hunting and apriorist reductionism.

For there *is*, unquestionably, a typological sequence through which human societies have evolved. Spencer is quite right that societies have become progressively more heterogeneous and complex through an increasing division and specialization of roles, and Marx is quite right that this process has involved modal shifts in which a novel set of coherently interrelated institutions comes to displace that out of which it has evolved. If both, from their different perspectives, over-stressed the primacy of economic institutions, practices, and roles, that can be remedied readily enough by conceding to the coercive and the ideological such autonomy as the evidence of the historical and ethnographic record requires; and if both were a little too apt to leave the impression that although there is no unilinear sequence there is a presumption of regularity to which the exceptions are to be explained away, that can be remedied by denying the presumption outright. The serious difficulty, once again, is the connection between sequence and progress. If there is a single quotation from Spencer which sums it up, it is where he says in *The Study of Sociology* that 'if there does exist an order among those structural and functional changes which societies pass through, knowledge of that order can scarcely fail to affect our judgements as to what is progressive and what is retrograde'[22]—to which the response must be that there is

classes of the community marked-off by rank, and sub-classes marked-off by special occupations, severally combine, and severally set up organs advocating their interests: the reason assigned being in all cases the same—the need for self-defence.'

[22] Op. cit., 64.

indeed such an order but judgements of what is progressive or retrograde have nothing to do with why that order has been what it has. If there is a single quotation from Marx which sums it up, it is where he says in a well-known passage in the Preface to the *Contribution to the Critique of Political Economy*, that 'in broad outlines we can designate the Asiatic, the ancient, the feudal, and the modern bourgeois modes of production as epochs in the progress of the economic formation of society. Bourgeois relations of production are the last antagonistic form of the social process of production'[23]—to which the reponse must be that there are indeed qualitatively different epochs but their progressiveness is in the eye of the beholder and if post-bourgeois modes of production are non-antagonistic then the concept of antagonism has been defined out of all possible recognition. It is, admittedly, true that there are difficulties to be overcome both in formulating a precise definition of the distinctive types (and sub-types) of societies which have evolved thus far and in identifying the precise point in the sequence at which the structural and functional change from one to another can be said to have occurred. But these difficulties are not arguments against the validity of the propositions which give rise to them. Whether or not feudalism is defined in terms of the centrality of the fusion of the institutions of benefice and vassalage, and whether or not the emergence of capitalism (or the 'bourgeois' mode) is dated prior to the large-scale use of wage-labour for industrial production organized for private profit, there is no doubt that the societies conventionally labelled 'feudal' and 'capitalist' are indeed distinctive and, moreover, that we now understand their structure and functions rather well. What is more, although there cannot yet be claimed to be a single definitive explanation of how the evolution of the latter out of the former first came about, we do now know a remarkable amount about the conjunction of ecological, demographic, and sociological causes which set in train the sequence of individual actions whose unintended consequence *à la* Spencer was a qualitative shift *à la* Marx.

Given this measure of broad agreement at the theoretical level,

---

[23] Marx and Engels, *Selected Works* (Moscow and London, 1951), i. 329 (translation slightly modified): the critical phrase in the original is '*als progressive Epochen der ökonomischen Gesellschaftsformation*'.

and given also the amount and quality of empirical research on human societies of all types which has been carried out in the hundred years since the publication of Volume I of Spencer's *The Principles of Sociology*, the most surprising lacuna—at least to me—is the lack of any attempt to answer systematically the question how many distinctive types of society are possible at any given evolutionary stage. If Spencer's institutional variables are many too many and Marx's modes of production many too few, why has no sociologist between then and now set out to produce a classification broad enough to embrace the major differences in political and ideological as well as economic institutions and narrow enough to exclude all the innumerable cultural differences which are irrelevant to the process of evolution from one to another mode of the distribution of power? If there are, indeed, clusters of dominant institutions which give terms like feudalism and capitalism their meaning; if there are discernible stages in the expansion of resources which make new modes of production, persuasion, and coercion possible; if there are nevertheless constraints which limit the logically possible range of institutional variation; and if, finally, variations in the structure of societies are necessarily related to the functions which their economic, ideological, and political institutions perform, then it must be possible to construct a manageable typology which generates comparisons and contrasts whose theoretical rationale will be equally acceptable to Spencerian and Marxian evolutionists alike. The reason for which it has not been done is, I suspect, not only because taxonomy is dismissed as 'butterfly-collecting' but at least partly because, yet again, evolution is assumed to be progress towards some necessary outcome. If the great irreversible changes initiated by the so-called industrial revolution are going sooner or later to bring about a convergence of all human societies in response to a common set of inexorable functional imperatives, then differences in their earlier organizational forms are only of marginal significance. But the notion that industrialization implies convergence is as palpably fallacious in its individualist, Spencerian version as in its collectivist, Marxian one. We can now see, as Spencer and Marx could not, that industrialization, like every major evolutionary shift, brings about, as I said earlier, both increasing and diminishing variation at the same time. If we survey the late twentieth-century world from China to Peru by way of Iran, Yugoslavia, Israel, South Africa, Sweden, Nigeria, and Brazil, we can hardly fail to be struck by the different ways in which the inescapable institutional transformations which industrialization brings about are modified in response to ideological and political pressure. Evolution, yes: new institutions, practices, and roles are visibly displacing old ones through a process of competitive selection. But progress, no: there is no necessary outcome, and if the state of the various industrial societies of the world is, after some specified interval, better than it was before, that is for you to decide in the light of whatever your personal values may happen to be.

To sum up, then: my theme has been that by learning from the mistakes of the two greatest nineteenth-century evolutionary sociologists we can, instead of losing the baby with the bathwater as too many of their critics have done, carry forward the task which in their different ways they rightly set themselves—the task, that is, of doing for the study of societies what Darwin did for the study of species. It is easy, in 1986, to ridicule the high confident Europocentric assumption that secularization and steam-engines would between them usher in for mankind a peaceable and prosperous millennium in which the outdated institutions of church and state would be dropped together, with cries of relief, down the *oubliette*. But once subtract from the concept of social evolution the presupposition of progress, and it is remarkable not only how much in Spencer and Marx alike can be jettisoned to advantage but also how much of enduring value remains. Even if giants have feet of clay, they can still support pygmies on their shoulders; and if there are two giants, and the pygmy puts half of his modest weight on one shoulder of each, he can stand firmly upright without fear that either will give way under him.

3

Darwinism and Behaviourism: Parallels and Intersections[1]

R. J. HERRNSTEIN

The doctrine of evolution by natural selection was in hot water from the start. Charles Darwin defensively called his voluminous first book on the topic, *The Origin of Species*, an abstract. Quite an abstract it was: hundreds of crowded pages about variation and inheritance of traits affecting survival, about prolific reproduction winnowed down by the hazards of life, about similar but not identical races and species scattered across the globe and in the fossil record, and about the creation of new life forms by artificial selection of domestic animals and plants. Since then, additional supporting facts from natural history have piled up mightily. And beyond those, we now know about the double helix—the actual mechanisms of inheritance at the molecular level. Evolution has, on a small scale, been brought into the laboratory and market-place. Yet, the doctrine remains in hot water in some circles. My theme here is that the contemporary troubles mainly concern behaviour. In addition, I will suggest that new discoveries about behaviour draw biology and psychology even closer together than they have seemed to be. The implications for social policy unequivocally support neither the original social Darwinians nor their critics.

Among the early critics of evolutionary thinking were socio-logists of the political left reacting against the evolutionary approach to social theory, such as Herbert Spencer's.[2] Spencer said that human society, like animal life in general, was a crucible

[1] An earlier version was read at the University of Oxford on 7 Nov. 1986 as a Herbert Spencer Lecture. I am grateful to the Board of Managers of the Herbert Spencer Fund for this opportunity to think again about the rich connections between evolutionary theory and psychology. Thanks also to my Harvard colleague, William Vaughan, jun., for permission to include here some charts that he and I have previously published elsewhere, along with some ideas that the two of us have developed in collaboration.

[2] See, for example, *Social Statistics* (New York, 1864).

for survival of the fittest, his vivid and lasting phrase. To Spencer, successful citizens were winning a biological contest no less than successful animals in nature, and were passing on their traits to future generations. In this way, man and beast evolve towards greater adaptability to their surroundings, which, for man, was society.

His critics argued that, rather than being a crucible for natural selection, society evolves culturally. Mankind's biological make-up was not changing as history unfolded, said the critics; what was changing were the lessons that each generation taught the next, as embodied in institutions and policies. The critics from the political left believed they understood the lessons well enough to create a better society. They wanted to remake man by changing society, rather than vice versa, which is what Spencer advocated.

Spencer and his social Darwinist followers believed that biology shapes human society. Reform, humanitarian though it may be, would pollute the gene pool by rescuing the weak from their biological destiny, namely extinction. Spencer's political sympathies were conservative and gradualistic; he favoured *laissez-faire* policies that allowed individuals to compete freely, so as to permit natural selection to forge gradually what he expected to be better biological adaptations to social life, such as altruism and community spirit.

The less patient or optimistic of his successors, thinking of the competition between races or states rather than individuals, were dissatisfied with *laissez-faire*ism, instead, they believed they found biological justifications for restrictive immigration, militarism, imperialism, and even war. The Darwinian themes of German Nazism, for example, are well known, but I need not go so far afield to find examples. Said Theodore Roosevelt in the closing days of the nineteenth century, when he was Governor of New York State, lately hero of the Rough Riders at San Juan Hill in Cuba, and a conscious social Darwinian:

If we stand idly by, if we seek merely swollen, slothful ease and ignoble peace, if we shrink from the hard contests where men must win at hazard of their lives and at the risk of all they hold dear, then the bolder and stronger peoples will pass us by, and will win for themselves the domination of the world.[3]

[3] This, and the other historical quotations to come, are taken from, and documented in, R. Hofstadter, *Social Darwinism in American Thought* (rev. edn., Boston, 1955).

Despite such examples, belief in evolution does not neatly divide conservatives from progressives, the right from the left. The leading American neo-conservative, Irving Kristol, suggested recently in the *New York Times* that, since the origin of species by natural selection was only a hypothesis rather than a proven fact, it should be taught as such in American schools. He finds current practice in the teaching of evolution to be too dogmatic and too anti-religious. He did not go so far as to propose that it share the curriculum with the creationist theory of Revd Jerry Falwell and other religious fundamentalists, which is approximately the same theory of special creation that Darwin believed he refuted over a century ago.[4] Professor Kristol's comments were reminiscent of Ronald Reagan's in his first presidential campaign and early in his administration, when he seemed closer to the religious right than his policies have proved to be. Spencer found in evolution support for conservative politics; some modern conservatives instead find a threat.

The objection to evolution from the right is as traditional as the objection from the left. From the beginning, evolution by natural selection greatly disturbed many clergyman and other guardians of morality. The design of the biological world—the exquisite fit between creatures and their environments—was long an argument for the existence of God—the so-called argument from perfect design. Only God himself was capable of creating perfect living forms. The Bible's story of creation plainly differed from Darwin's. Evolution by natural selection gave credit for the design to blind natural forces. In scripture, creation was purposive; in Darwin, it was just higgledy-piggledy, to echo John Herschel, one of his early critics. Mankind's divine creation seemed to be undermined by what Darwin called the 'hairy quadruped' in our ancestry. Those were all grounds for criticism, but probably most objectionable about natural selection was its threat to morality.

To erode religious doctrine was, it was feared, to erode religion's authority over good and evil. If religion is wrong about the origin of the world, wrong about the origin of species, and wrong about the origin of mankind itself, can we be sure it is right about how we should behave? And if we cannot be sure, how much will the balance shift from good to evil conduct in a post-

[4] 30 Sept. 1986 (A 35).

Darwinian world? It is this concern, bolstered by clear and alarming evidence of the deterioration of morals in modern society, that motivates Professor Kristol and the President, not to mention the fundamentalists with whom they are allied. Concerns about individual conduct freed of the constraints of religious morality are transformed into doubts about the doctrine of evolution as science.

Finally, as a contemporary example of the objection from the left, consider the interesting case of palaeobiologist Steven J. Gould, currently the most celebrated and eloquent antagonist of the creationists, defending evolutionary doctrine from those who would demote it to a mere hypothesis, or less, in American schools. More, not less, evolution taught in the schools is what this biologist wants. Curiously, Professor Gould is also the most celebrated and eloquent American antagonist of modern socio-biology, the notion that biology plays something like the role Herbert Spencer attributed to it in the shaping of society. How can a person defend evolution against creationism while repudiating it for sociobiology? Echoing some of the early critics from the left, Professor Gould accepts the doctrine of evolution in relation to animals, but denies its relevance to the human condition.[5] Professor Gould might agree with Lester Ward, an early anti-Spencerian, who wrote:

As far as ... native capacity ... [is] concerned, those swarming, spawning millions, the bottom layer of society, the proletariat, the working class, the 'hewers of wood and drawers of water,' nay, even the denizens of the slums— ... all these are by nature the peers of the boasted 'aristocracy of brains' that now dominates society and looks down upon them, and the equals in all but privilege of the most enlightened teachers of eugenics.[6]

From the left, anti-Darwinism argues that the human capacity to learn from experience substitutes cultural evolution for bio-logical. The human inequalities that we observe must, the argument goes on, result from unequal treatment by society, not unequal genes. If, for example, we discover that the fats in red meat are unhealthy, we need not wait until mutation provides a gene for disliking red meat, which natural selection would then

[5] See, for example, *The Mismeasure of Man* (New York, 1981).
[6] 'Social Darwinism', *American Journal of Sociology*, 12 (1907), 407.

favour in the harsh and slow process of differential survival. We need simply to learn the facts and behave accordingly. Nor would it seem plausible to these advocates of reform or revolution that those who still eat beef and those who have already stopped differ in this regard because of a difference in genes. Those who have stopped, the critics would argue, have had some cultural, not biological, advantage. For mankind, they believe, unlike for animals, knowledge itself, acquired not innate, can rapidly and almost painlessly change behaviour and the world. In the words of journalist Walter Lippman, anti-social-Darwinian and early advocate for an American welfare state:

We can no longer treat life as something that has trickled down to us. We have to deal with it deliberately, devise its social organization, alter its tools, formulate its method, educate and control it. In endless ways we put intention where custom has reigned. We break up routines, make decisions, choose our ends, select means.[7]

Critics from the left and right, antagonists on political battle-fields, nevertheless agree that human behaviour, or some signific-ant part of it, can be hammered into almost any shape on the anvil of circumstance and experience. To the left, malleable human behaviour justifies reform or revolution. If some people are poor and others wealthy, smart or stupid, criminal or law-abiding, it is because of environment, not genes. To change them, change their circumstances or abolish the social institutions that the critics blame for the problems, such as unrestrained capitalism or monopoly or differential opportunities for education.

To critics of the right, the issue is more narrowly focused on morality, but otherwise the same. People behave the way they are taught to behave, these critics believe. If evolutionary doctrine threatens religious teachings, it threatens moral conduct. Religion is seen as an indispensable instrument of indoctrination for virtue. The threat of its loss justifies denying the evidence for evolution, or at least withholding it from the general public.

Those, on the other hand, who believe that human behaviour is significantly constrained by genes—social Darwinians of the past, sociobiologists today—may be less optimistic about reform or revolution. They are also less worried about the moral hazards of

---

[7] *Drift and Mastery* (New York, 1914), p. 267.

biological science than about the dangers of biological ignorance. Evolutionary theory says that behaviour can in principle express inherited adaptations no less than feathers or fins. We do not expect bodily structure to be as malleable as iron on an anvil, and we should not expect behaviour to be so either. Some critics accept this reservation about the limits of behavioural change in principle, but deny its practical relevance to human society.

Nature versus nurture in regard to behaviour is the last great evolutionary controversy. All the other arguments swirling around Darwinism in the nineteenth century have largely ceased. If there are still quarrels about vitalism versus mechanism, purpose versus higgledy-piggledy, Platonic essentialism versus fundamental variation, they are academic. Only arguments about the sources and consequences of behaviour still churn with enough energy to spill over into public debate.

Controversy surrounds, for example, the accumulating evidence for heritable variations in intelligence or personality. We know something about those things now because of psychological measurement and quantitative genetics, neither of which was available in Spencer's day, but which developed largely inspired by Darwinism. Since social behaviour often depends on intelligence or personality, it is possible that genes play a significant role in how social institutions impinge on individuals, and even in the possible forms that social institutions take. Cultural evolution may not displace biological evolution; rather, both could go on, with each constraining the other.

Significant as heritable variations in individual human psychology are, they seem to be increasingly well known and need no review here, while another aspect of Darwinism in relation to behaviour gets little attention. Darwinism, in replacing teleology with natural selection, provides a model that applies to individual behaviour as well. The ramifications of this model may alter conceptions of individual behaviour no less than the doctrine of natural selection alters conceptions of living forms.

Ontogenetic Parallel to Natural Selection

Variation and selection are the two essential components of evolution. Individual members of a species may vary in size, shape,

biochemical composition, nutritional needs, behavioural tendencies, and so on and on. Some of the variation arises in the
varying experiences of the individuals; but some of it, due instead
to differences in their genes, would be heritable, passing from
generation to generation via the genotype. When the heritable
variation affects reproduction and survival of offspring, selection
takes place. Over generations, the genotype shifts toward those
with the greatest reproductive success. Less successful genotypes
become rarer, possibly extinct. What looks to the untrained eye
like a purposive unfolding of ever-more-adaptive forms seems, in
the light of evolutionary theory, to be just mechanistic selection.
Evolutionary theory thus naturalizes purpose. Purpose becomes
part of the biological world, rather than being outside it as divine
will or a mystical life force. If an evolutionist speaks of purpose in
his subject, it is only as a shorthand for natural selection itself, and
nothing more. More likely, he prefers not to mention purpose at
all, for evolutionary theory has displaced it.

Except, that is, when he, or most everyone else, speaks of
behaviour. Then purpose remains unabashedly in the vocabulary.
People rationalize acts by their goals. People feel sure they are
moved by their purposes, at least sometimes. The will is held to
control our behaviour, just as, in the eyes of the pre-Darwinian
observer, the creator controls his creation. However, the central
theory of modern behaviourism displaces purpose in the analysis
of individual behaviour by a formal structure that is the
ontogenetic parallel to natural selection in evolution. Most of the
quarrels about social Darwinism and sociobiology arise in
extensions of the concept of biological *variation* to the human
sphere, but there is also a way to extend the concept of *selection*.

I refer to reinforcement theory. First, I describe the theory
qualitatively, then I show that its mathematical structure is
identical with that of evolution by natural selection, to the extent
that those structures are known at the present time. Our world
view is, I will suggest, changed more profoundly, and, for many
people, more disturbingly, by a naturalistic account of behaviour
than it was by a naturalistic account of the origin of species.

Reinforcement theory, stated qualitatively, is obvious and
ancient. In the nineteenth century, Herbert Spencer, Alexander
Bain, Conwy Lloyd Morgan, and many others, noted that

behaviour is shaped by its consequences.[8] In the eighteenth century, Jeremy Bentham had made it the keystone of the utilitarian conception of human behaviour.[9] It would not be hard to trace the central idea in the theory back to the earliest hypotheses about behaviour. The idea is that a sample of an organism's behaviour is strengthened when it produces a reward and weakened when it produces a punishment. Picture a hungry rat in a Skinner box, named after the leading modern exponent of reinforcement theory. The rat depresses a lever and earns a food pellet. From such a contingency between behaviour and consequence, the lever-pressing behaviour increases in strength; the rat spends more of its time pressing the lever when it is hungry. Other examples can be multiplied without limit.

The parallel to evolution by natural selection, again speaking qualitatively, is also obvious.[10] Reinforcement, like evolution, is a matter of variation and selection. The rat's behaviour varies from moment to moment; reinforcement selects some forms of movement over others. What may seem to the untrained eye to be a rat with a goal, pursuing its purposes by pressing a lever for rat food, is better viewed, according to reinforcement theory, as a rat whose behaviour has been selectively shaped by the consequences of past action.

Let us consider the analogy further. In evolution, the phenotype is directly observed, but it is the unobserved genotype that is selected. A given genotype may express itself in different phenotypes, depending upon the environment. Identical twins, sharing the same genotype, are not phenotypically identical: two bodies cannot simultaneously occupy the same space, so twins cannot develop in precisely the same environments. Reinforcement theory contains a comparable subtlety. We observe movement, but something more elusive is strengthened, more like a class of potential movements having a common effect on the environment. It is not a fixed pattern of muscular contractions that is strengthened when lever-pressing is reinforced, but a broader class of actions having in common the pressing of the lever. The rat may

[8] See R. Boakes, *From Darwinism to Behaviourism: Psychology and the Minds of Animals* (Cambridge, 1984).

[9] *An Introduction to the Principles of Morals and Legislation* (London, 1789).

[10] R. J. Herrnstein, 'Will', *Proceedings of the American Philosophical Society*, 108 (1964), 455–8.

use different paws on different occasions, or even his muzzle. Different environments or different orientations of the rat's own body call forth different movements, when a given class is strengthened by reinforcement. The phenotype is to the genotype as a particular movement is to a class of actions.

The analogy goes on. In evolution, reproductive fitness is more subtle than the layman's, or the early biologist's, intuition about it. 'Survival of the fittest' seems to say something about the absolute value of the selected hereditary traits. The layman pictures the fittest creature as a large, robust, fertile one glowing with health. Often they are just those things, but not necessarily. It need not be large size, or great strength, or extra intelligence, or anything else in particular, that confers advantage on an individual. Reproductive fitness is, rather, whatever the given circumstances of life say it is. It may be an advantage to be small, or weak, or not so smart.

Likewise, a reward may benefit a creature objectively, and a punisher may harm it, but not necessarily. A naïve trust in the order of things may lead us to believe that rewards are the things animals should seek, not just do seek; punishments, the things they should avoid, not just do avoid. Doing what comes naturally is held to be a good idea. But the reinforcers for a species reflect an evolutionary history that may or may not bear on current circumstances. We are, for example, more reinforced by sweet tastes and by expressing anger than we ought to be in a world with abundant sugar and potent weapons. As a result we live with the risk of obesity and of destroying each other. We are too much punished by the sensations of physical exertion than is appropriate for modern sedentary life. We may become flabby or vulnerable to stress. Addictive substances are reinforcers gone wild. Reinforcers are, in short, whatever stimuli have the strengthening or weakening effects on behaviour, and not necessarily anything more. Reinforcers and punishers may be beneficial or lethal: overeating, overbreeding, overkilling, and underexercising are among the many dangers of reward and punishment in modern life.

True teleology or purpose, guided deliberately, has been replaced by natural selection, and, according to modern behaviourism, by reinforcement. Will, divine or individual, ceases to be in charge; nothing beyond the selective process itself is in charge, just as nothing beyond the laws of matter in motion is in charge of placing the grains of sand on a beach. This is a familiar theme for the origin of species, but not for the analysis of individual behaviour. I am suggesting, in short, that reinforcement replaces will the same way as natural selection replaces teleology. It is not an easy suggestion to accept; giving up free will may be harder than giving up divine purpose.

Besides the metaphysical shock of replacing teleology with mechanism, there is the further shock of realizing that we are unlikely to be as well taken care of as we thought. Mechanism does not work as well as divine or individual will in achieving a goal. Natural selection in biology does not create perfect designs, refuting the central premise of the old theological argument from design. Creatures are selected by the more modest and uncertain criterion of comparative advantage. Are they better than the competition is evolution's question, not are they perfect.

A similarly modest and uncertain criterion shapes individual behaviour, as I will try to show. The ontogenetic parallel to the theological argument from perfect design is the model of rational choice at the heart of neoclassical economics. Modern economics depicts human behaviour as approximating rationality; its magnificent mathematical structure formalizes an intuition that each person, left to his own devices and disregarding errors of calculation and the like, will tend to optimize utility, the economist's word for reinforcement. Lately, the theory of rational choice has been adopted by other behavioural disciplines, such as the optimal foraging models of modern biology. It is this notion of rationality that I hold to be as false a model of individual behaviour as special creation and perfect design are of the origin of species.

A Formal Theory of Selection

Let us first consider how modern evolutionary theory has formalized natural selection, essentially as a problem in game theory. I use an example first discussed by John Maynard Smith.[11] Imagine a hypothetical species whose behaviour is much influenced

[11] See, for example, *Evolution and the Theory of Games* (Cambridge, 1982).

by which of two genes is present at a particular location: it is convenient to call one the 'hawk' gene, the other, the 'dove' gene. The hawk gene causes its bearer to fight every time it encounters another member of the species until one wins and the other loses. The dove gene causes its bearer to flee if it encounters a hawk, thereby losing the contest but suffering no injury. When a dove meets a dove, both posture for a while; one wins the contest and the other loses, but neither one is injured. In a fight between hawks, only the loser is assumed to be injured. To make the example relevant to natural selection, we must now assign values, measured in units of reproductive fitness, to the various outcomes.[12] I will assume that the winner of a fight earns 50 fitness points; the loser, 0 points. An injury costs 100 points, and the posturing that doves engage in costs 10 points.

If there are only hawk genes, then every encounter results in injury to one of the animals, in addition to the one victor and one loser. The average fitness outcome of a fight is then 50 (for the victor) plus 0 (for the loser) minus 100 (for the injury to the loser) divided by 2 (since the outcome is spread over the two animals fighting), which equals −25 fitness points. If a dove mutation appears, it will meet only hawks. An encounter with a hawk will earn the dove 0 points, since it loses but immediately runs away. In a population of virtually all hawks, then, a solitary dove has the biological advantage, all else equal, since its 0 points is better than an average hawk's −25 points.

But in a population of virtually all doves, the tables are turned. Fights between doves have an average fitness of 15 points: 50 to the winner plus 0 to the loser minus 20 for posturing by each of them, divided by 2. A hawk mutation, by winning all contests with doves, would earn 50 points and have the biological advantage. In short, dove populations would be infiltrated by hawk genes, and hawk populations would be infiltrated by dove genes, due to the process of natural selection. At some mixture of hawk and dove genes, an equilibrium is struck. Figure 1[13] illustrates the equilibrium point for hawk and dove genes, given the fitness values I assumed.

Along the x-axis are proportions of dove genes in a population;

[12] W. Vaughan jun. and R. J. Herrnstein, *Stability, Melioration, and Natural Selection*, in L. Green and J. Kagel (eds.), *Behavioral Economics*, vol. i (Norwood, NJ, 1987).
[13] Fig. 5, ibid.

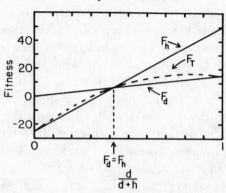

FIG. 1. Fitness of 'hawk' gene, F_h, and 'dove' gene, F_d, as a function of the proportion of dove genes in the gene pool. The evolutionarily stable strategy, where hawk and dove genes have equal fitness, is at about 42% dove genes. Maximum overall fitness is at about 83% dove genes, which is the highest point on the dashed curve plotting the average fitness for the population as a whole.

along the y-axis, the fitnesses earned by hawks or doves for the given proportions. The extremes are the situations I have already mentioned: −25 fitness points for hawks in a pure hawk population; 0 for doves in a virtually pure hawk population; 15 for doves in a pure dove population; and 50 for hawks in a virtually pure dove population. The solid lines trace the fitness points for hawks and doves given all mixtures of the two genotypes.

When the two genotypes have equal fitness, at about 42% doves and 58% hawks, equilibrium is reached. A change from this point in either direction would be self-defeating. An increase in doves would lower the doves' fitness relative to the hawks'; and likewise for the hawks. A decrease in doves would increase their fitness relative to hawks, and again likewise for hawks. Thus, at precisely this mixture, each genotype is non-invadable, hence stable. Biologists call such a mixture of genotypes an evolutionarily stable strategy, or ESS.

Maynard Smith's example is significant because the ESS may be suboptimal for the species as a whole. At the ESS, the average animal in the species has a fitness of 6¼, which is the y-axis value at the point of equilibrium; in contrast, if all of them were doves, the average fitness would be 15. The dashed curve traces the

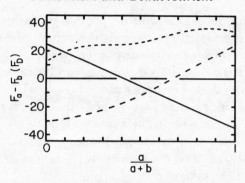

FIG. 2. Three hypothetical ways in which two forms, *a* and *b*, may have the difference between their fitnesses, F_D (i.e. F_a-F_b), depend on the proportion of *a*s in the population. The downward-sloping solid line, which describes the hawk–dove example in Figure 1, produces an evolutionarily stable strategy at about .42 of the *a* form. The upper dashed curve would result in a total exclusion of *b* by *a*. The lower dashed curve would produce either all *a* or all *b*, depending on initial conditions.

weighted average of the fitness of hawks and doves taken together. It shows, in other words, how the species as a whole should fare. The point of maximum fitness is the highest point on the dashed curve, at about 83% doves and 17% hawks. Although optimal, this mixture of genotypes is unstable, for hawks are doing considerably better than doves, and the population must therefore shift toward the ESS of 42% doves and 58% hawks.

The failure of natural selection to produce the true optimum for the species shows concretely why teleology, if it existed, would work better than higgledy-piggledy. An external, benevolent hand should, given the variables, guide this species toward 83% doves and 17% hawks. The 'invisible hand' of evolution by natural selection will fail to do so well. Indeed, imagine that our hypothetical species, with its stable mixture of 42% doves and an average fitness of 6¼ is in mortal competition with a species whose average fitness is, say, 10. The dove-hawk species will become extinct, even though it could have prevailed if it had kept its dove proportion above about 55%.

The ESS is defined as the point at which the fitnesses for the competing forms are equal and any deviation from equality is self-correcting. Figure 2[14] illustrates three simple ways two forms may

[14] Fig. 6 in Vaughan and Herrnstein.

interact. Along the x-axis is the proportion of form *a*; along the y-axis is the fitness of form *a* minus the fitness of form *b* for the corresponding proportion of form *a*. The downward-sloping solid line resembles the dove-hawk example, producing an ESS at some mixture of *a*s and *b*s. The upper dashed curve will produce a pure *a* population, since the *a*s have higher fitness at all mixtures. The lower dashed curve will produce either all *a*s or all *b*s, depending upon initial conditions. If the population gets to the right of the equality point (i.e., 0 on the y-axis), *a*s take over; to the left of the equality point, *b*s take over.

Evolution by natural selection fails to optimize in general because, as Figure 2 illustrates, the process that guides it has no representation within it of the overall fitness of the species. The figure does not tell us where the optimum would have been. Figure 3 (Figure 7 in Vaughan and Herrnstein) should make this point clearer. Each of the three panels in it depicts competition between two genes, *a* and *b*. The x-axis again shows the proportion of *a*s in the population. The fitness averaged over the species as a whole is again the dashed curve. The differences between the fitness of *a* and the fitness of *b* is shown by the curve labelled F_D. In all three examples, natural selection would yield a pure *a* population, because *a* has greater fitness than *b* at all mixtures of the two genes: F_D is above zero under all conditions. However, the overall fitness is maximized at three different mixtures: in A, maximum fitness is at a pure *b* population; in B, it is at a pure *a* population; in C, it is at about 65% *a*s. As a dynamic process, natural selection, as we understand it, is virtually blind to overall fitness of a species. This says that a species may evolve as more fit structures drive out less fit ones, but there is no guarantee that the overall fitness of the species' gene pool is moving towards better adaptation. Natural selection adapts locally, not globally. Case A is, in fact, the gloomy reverse: as *a* comes to predominate, the species as a whole goes downhill.

Many writers about evolution and behaviour, even contemporary ones, say something to the effect that, inasmuch as evolution is an optimizing process, it should select behavioural laws that are themselves optimal in promoting reproductive fitness. Biologists assume that behaviour must be optimal or nearly so, then try to figure out what the creature's behaviour has

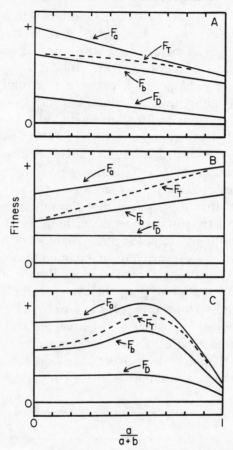

FIG. 3. Each panel shows, as a function of the proportion of form *a*, the fitness of *a* (F_a), the fitness of *b* (F_b), the average overall fitness (F_T), and the difference in fitness between *a* and *b* (F_D). In each panel, *b* would be totally excluded by *a*. A pure form-*a* population maximizes fitness for panel B, but minimizes it for panels A and C.

optimized, just as economists do in applying their notion of rational choice. If a biologist can plausibly infer some optimal outcome for some observed behaviour, he and his colleagues are likely to consider the behaviour explained. But, given Maynard Smith's example and others like it showing that natural selection does not necessarily optimize, the logic of the argument unravels. At best, evolution would yield behavioural laws that are non-invadable, which is not the same as optimal.

It has often been shown in the laboratory that animals and
humans distribute their choices so that a more profitable
alternative receives an increasing share of the choices—'profitable'
measured in terms of reinforcement. Thus, if a pigeon confronts
two choices, one yielding more reinforcement than the other, its
behaviour will shift towards the better choice. Indeed, the
operational meaning of an alternative with 'more reinforcement' is
that, given a choice between two or more alternatives, there is a
shift in behaviour towards that alternative. Figure 4[15] diagrams
this utilitarian outcome for two mutually exclusive and exhaustive
alternatives. Let us say a pigeon has a certain amount of time to
spend on alternative a or b, and that the value he receives from
either, measured by reinforcements per hour, is, on the average,
constant. A human example, often simulated for laboratory
pigeons, would be a slot-machine in a gambling casino: each coin
played has a certain probability of winning. Because the probability
is unaffected by the rate of play, the average number of wins per
unit time is fixed, so long as we play at a steady rate. Our earnings
per unit time, whether positive or negative, are also fixed for each
of the alternatives. Figure 4 shows two competing alternatives, one
more lucrative than the other. Overall earnings now depend on the
allocation to the two alternatives. Given sufficient experience, a

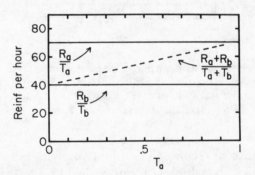

FIG. 4. Rate of reinforcement as a function of the proportion of time, T_a,
invested in one of two mutually exclusive, exhaustive, response alternatives, a
and b ($T_a + T_b = 1$). The rate of reinforcement obtained from either alternative,
R_a/T_a and R_b/T_b, is independent of the proportion invested in it. Overall reinforcement rate, shown by the dashed line, is maximized by exclusive choice of a.

[15] Adapted from Fig. 3, ibid.

pigeon's or man's choices will shift to *a* exclusively. The overall rate of reinforcement, represented by the dashed line, is here maximized. Behaviour would appear to be rational by the criterion of optimizing reinforcement.

Now let us consider a different sort of situation, but still one that is within the realm of everyday experience. In tennis, a player must decide, when his opponent has come to the net, whether to respond with a lob or a passing shot—whether, in other words, to try to hit the ball to the side or above the reach of the opponent at the net. Taking into account the element of surprise, the effectiveness of each shot depends inversely on how often it has been used in the preceding play. (It also depends on other things, such as the skill of the player, on his court position, and on similar features of the opponent, but those complications can be ignored here with no loss in generality.)

We can attach values, measured as points per shot, to the extremes, as in the hawk-dove example. Suppose, for example, that if the player has been hitting nothing but passing shots when his opponent is at the net, then a lob would earn .9 points per shot. The surprise lob is, in other words, almost infallible. But if the player has been hitting only lobs, then the opponent will anticipate it, and points per shot would be only .1. Passing shots, too, depend on how frequently resorted to, but not so much, because anticipation by the opponent has less effect on the success of this shot. If nothing but lobs are being hit, a passing shot is worth, say, .4 points; if nothing but passing shots are being hit, its value is .3 points per shot. These values are based on a notion that the opponent can somewhat neutralize a passing shot by guessing which side it is going towards, and can do so a bit better if he can anticipate that it is going to be a passing shot of some sort.

Figure 5 plots the values at the extremes and connects them by straight lines, which again is a simplifying assumption that involves no loss in generality. The horizontal axis is the proportion of lobs at a given point in the game (leaving aside the issue of how or how accurately this proportion is estimated by the player), and the two lines give the points per shot for lobs and passing shots.

Intuition, and the observed behaviour in many formally comparable laboratory experiments on animals and humans, suggest that the player comes into equilibrium where the two lines

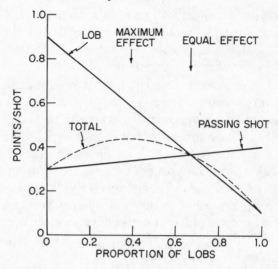

Fig. 5. Hypothetical points per shot for passing shots and lobs as a function of the proportion of lobs, based on the idea that anticipation reduces the effectiveness of lobs more than it does passing shots. At .667 lobs, the two shots have equal effectiveness. The overall effectiveness of the two kinds of shot taken together is given by the dashed curve, which passes through a maximum at .389 lobs.

cross. At all points except at the intersection of the two lines, one shot or the other is more successful; the tendency would be to use the more successful shot more often. At the intersection, the player is hitting lobs .667 of the time (and passing shots .333 of the time), and the average value of each is about .366 points per shot. If he shifts toward more lobs, then passing shots are better, and vice versa if he shifts towards more passing shots. The intersection point is the ontogenetic equivalent of an ESS, a stable allocation of behaviour based on a comparison of reinforcements received from the competing alternatives.

The intersection also violates rationality, for the player would earn the largest number of points if he hit lobs .389 of the time and passing shots otherwise. At that mixture, he would earn, overall, .436 points per shot. The dashed curve shows the overall outcome of all mixtures of lobs and passing shots. According to the utility maximization theory of modern economics, players should stabilize at this point-maximizing strategy. (I freely assume here that tennis

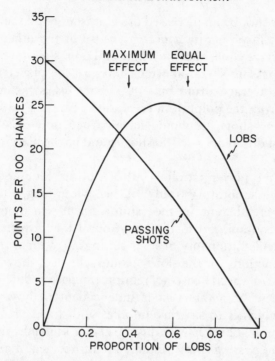

FIG. 6. Given 100 occasions for choosing between lobs and passing shots, these curves plot the numbers of points earned by each kind of shot as a function of the proportion of lobs, assuming the effectiveness of each kind of shot follows the functions in Figure 5.

players try to maximize their chances of winning, which roughly translates into scoring more points.)

Figure 6 shows in graphical terms how neoclassical economics approaches the question of behavioural allocation. Assume that the player has 100 occasions to choose between lobs and passing shots. The straight lines in Figure 5 provide the basis for calculating the results of all mixtures of choices; the outcomes are in Figure 6. The downward sloping curve shows the points earned by passing shots as a function of the proportion of lobs. The inverted U curve is the points earned by lobs, also as a function of the proportion of lobs. Economic theory holds that a player will tend to find that point on the two curves where the slope of one equals the negative of the slope of the other, for this is where the marginals are equal. At this point, the additional points earned from either shot by an increased use of it is more than balanced off by the loss in points by a decreased use of the other shot. The marginals have equal absolute value and opposite sign at .389 lobs and .611 passing shots, at which mixture the player earns 43.6 points (on average) from his 100 shots; this is the maximizing strategy. From the point of intersection in Figure 5, .667 lobs and .333 passing shots, he would have earned only 36.6 points. A seven-point difference per 100 shots could be the margin of victory in a match.

Why is the player rational with a slot-machine and not with tennis? The laboratory study of behaviour suggests that the difference is inherent in the reinforcement schedules that the situations contain, not in the behavioural law governing choice. The law yields rationality in some settings, irrationality in others. The flat functions for the slot-machine schedule, shown in Figure 4, comprise a special case—a reinforcement schedule that yields reinforcement at a rate that is independent of how often it is selected (which is to say, it reinforces with fixed probability). It should soon be obvious why this particular schedule, unlike those for lobs and passing shots, results in reinforcement maximization, and why reinforcement maximization, or rationality, is not a general law of behaviour.

Melioration, not Maximization

Figure 7[16] shows four examples of reinforcement schedules that are designed to see whether behaviour truly maximizes reinforcement, as rational choice theory says it does. In each case, alternative *a* is pitted against alternative *b*; what varies is how reinforcement varies with the allocation to one or another choice. The x-axis is always allocation to *a*; allocation to *b* is 1 minus that proportion. At zero, the subject is choosing *b* exclusively; at 1.0, *a* exclusively; at .5, *a* and *b* are being chosen equally. The solid lines show reinforcement for *a* and *b* at each allocation between the two choices.

In panel A, reinforcement for *a* is always a constant amount

---

[16] Fig. 4 in Vaughan and Herrnstein.

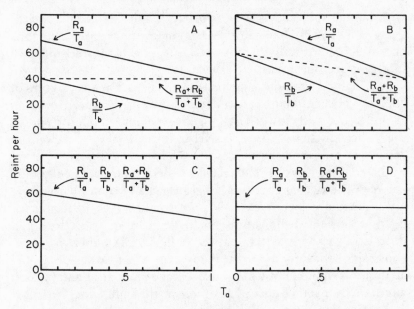

FIG. 7. Four schedules of reinforcement involving two mutually exclusive response alternatives, *a* and *b*, showing, in each panel, the reinforcement rate for each alternative and the overall rate of reinforcement as functions of the proportion of time allocated to *a*. A: Reinforcement rate for *a* is a constant amount larger than that for *b* at all allocations; overall reinforcement rate is independent of allocation. B: Reinforcement rate for *a* is again a constant amount larger than that for *b* at all allocations, but overall reinforcement rate is maximized when *b* totally excludes *a*. C: Reinforcement rate for *a* and *b* are equal and overall rate of reinforcement is maximized with exclusive choice of *b*. D: Reinforcement rate for *a*, *b*, and overall are all equal and independent of allocation.

better than for *b*. The overall reinforcement rate, given again by the dashed line, is horizontal, saying that the subject will earn the same overall pay no matter how it allocates choices between *a* or *b*. Pigeons working for feed in situations like this come to choose *a* exclusively. If they were rational reinforcement maximizers, they should be indifferent between *a* and *b*. In B, the only difference is that the two reinforcement lines have been tilted clockwise, so that overall pay is maximized by exclusive choice of *b*, even though, at any allocation, *a* earns a constant amount more than *b*. This is a queer situation because local and global adaptation are in direct conflict. Approximations to this decisive case have been examined with human subjects, as well as pigeon.[17] The result makes clear that reinforcement, like natural selection, operates locally. Subjects choose *a* preferentially, almost exclusively, at a considerable cost in pigeon feed or dollars.

In C, *a* and *b* earn the same reinforcement rate at any mixture of choices between them. Overall reinforcement, however, declines as preference shifts toward *a*. If pigeons maximized reinforcement, they should choose *b* exclusively; instead, they develop no consistent preference at all, despite weeks or months of exposure to the contingencies of reinforcement. Behaviour drifts willy-nilly between the two alternatives[18] because the two alternatives are always equally lucrative. This is a condition of neutral non-equilibrium, the behavioural analogue to a body in free fall, unconstrained by gravitational force. Finally, D similarly fails to constrain behaviour,[19] as a theory of either local or global adaptation would imply, inasmuch as each alternative and the two alternatives together always reinforce at the same rate.

A rational choice model of behaviour says that behaviour would be strongly controlled by the reinforcement contingencies in B and C, pushing choice toward *b*, and not controlled by those in A and D. A local mechanism, like that of natural selection in evolution, implies strong control over behaviour in A and B, pushing choice toward *a*, and no control in C and D. Abundant evidence, mostly from laboratory animals like rats and pigeons, unequivocally favours the local mechanism, which my colleagues and I have called melioration.

According to the principle of melioration, behavioural alternatives compete for the organism's time, just as genes compete for representation in the gene pool. The allocation of behaviour shifts towards alternatives with higher average rates of reinforcement. If an alternative always earns the highest rate of reinforcement, independent of how much time is being allocated to it, then it crowds out all other alternatives. The gambler's slot-machine, illustrated in Figure 4, is such a situation—other things equal, the

---

[17] Vaughan, 'Melioration, Matching, and Maximization', *Journal of the Experimental Analysis of Behavior*, 36 (1981), 141–9; Herrnstein, *A Behavioral Alternative to Utility Maximization* in S. Maitol (ed.), *Applied Behavioral Economics* (London, in press).

[18] Herrnstein and Vaughan, *Melioration and Behavioral Allocation*, in J. E. R. Staddon (ed.), *Limits of Action: The Allocation of Individual Behavior* (New York, 1980).

[19] Ibid.

alternative with the highest probability of winning will exclude all other alternatives. The tennis example is not, for the reinforcement for lobs and passing shots depends on how often they are used. Given the nature of the reinforcement schedules, melioration sometimes yields reinforcement maximization, as for slot machines, and sometimes not, as for lobs and passing shots. Behaviour in either sort of environment appears to be controlled by melioration.

Many laboratory experiments support melioration, rather than maximization, as the underlying law of animal behaviour.[20] Small amounts of data from human subjects, and from animals in natural settings, similarly favour melioration. The example of suboptimal ESSs shown earlier are reproduced in individual behaviour—at equilibrium, an organism's behaviour is in balance so that each alternative earns the same average pay, even at the cost of large losses in overall reinforcement. Individual teleology falls as short, and in the same way, as teleology in the origin of species.

Game theorists are well aware of the hazards of a locally adaptive process. A spectator at a football match rises to get a better view; soon everyone is standing and the individual advantages have been obliterated. A standing crowd sees the

[20] See, for example, Herrnstein, *Melioration as Behavioral Dynamism*, in M. L. Commons, R. J. Herrnstein, and H. Rachlin (eds.), *Quantitative Analyses of Behavior*, vol. ii, *Matching and Maximizing Accounts* (Cambridge, Mass., 1982); id., *A Behavioral Alternative to Utility Maximization*; G. M. Heyman and R. J. Herrnstein, 'More on Concurrent Interval-Ratio Schedules: A Replication and Review', *Journal of the Experimental Analysis of Behavior*, 46 (1986), 331–51. That laboratory experiments support melioration is, at any rate, my conclusion, and that of a number of other workers. See A. I. Houston, 'The Matching Law Applies to Wagtails' Foraging in the Wild', *Journal of the Experimental Analysis of Behavior*, 45 (1986), 15–18; D. Prelec, 'Matching, Maximizing, and the Hyperbolic Reinforcement Feedback Function', *Psychological Review*, 89 (1982), 189–230; id., *The Empirical Claims of Maximization Theory: A Reply to Rachlin and to Kagel, Battalio, and Green*, ibid. 90 (1983), 385–9; Vaughan, 'Choice: A Local Analysis', *Journal of the Experimental Analysis of Behavior*, 43 (1985), 383–405; B. A. Williams, *Reinforcement, Choice, and Response Strength*, in R. C. Atkinson, R. J. Herrnstein, G. Lindzey, and R. D. Luce (eds.), *Steven's Handbook of Experimental Psychology* (New York, 1988). Other workers favour one or another form of maximization theory, with no obvious consensus among them about the form of maximization or convergence of their theories. See H. Rachlin, R. Battalio, J. Kagel, and L. Green, 'Maximization Theory in Behavioral Psychology', *The Behavioral and Brain Sciences*, 4 (1981), 371–417; A. Silberberg, B. Hamilton, J. M. Ziriax, and J. Casey, 'The Structure of Choice', *Journal of Experimental Psychology: Animal Behavior Processes*, 4 (1978), 368–98; S. R. Hursh, 'The Economics of Daily Consumption Controlling Food- and Water-Reinforced Responding', *Journal of the Experimental Analysis of Behavior*, 29 (1978), 475–91; J. E. R. Staddon, J. M. Hinson, and R. Kram, 'Optimal Choice', ibid. 35 (1981), 397–412.

match no better than a sitting crowd. One factory reduces costs by cutting corners in the control of pollutants; when everyone does it the costs far exceed the aggregate individual costs of preventing pollution. Each fisherman or lumberman, left to his own devices, takes as much of his goods as is profitable for him. Unregulated, the result may be bankruptcy for all. The ghost of the extinct passenger pigeon offers mute testimony to the dangers of group behaviour, in this case, the behaviour of hunters each maximizing his take.

Game theorists tell us that collective action must be externally regulated to avoid those dangers. A standing spectator evokes the raucous disapproval of the folks behind him. Polluters and greedy fishermen are fined. The hawk-dove species would be well advised to monitor and regulate its proportion of each gene, if it could. Now we must add individual behaviour to the list of non-rational systems. The study of reinforcement tells us that individual behaviour is a collective activity in the same sense. An individual's behaviour is composed of competing alternatives, which vie for the individual's time. The more lucrative alternatives crowd out the less lucrative ones. The process may produce optimal benefits, but may not, depending upon the contingencies of reinforcement. The local adaptation called melioration may be suboptimal. It may be not only suboptimal, but as disastrous for the individual as polluting the environment is for the community, and need external regulation as much.

It is no great discovery that some individuals under some circumstances behave irrationally. Nor do I suggest that the particular dynamic process I have described explains all cases of human self-destructiveness. Behaviour breaks down when there is disease or derangement. Irrationality may arise in ignorance. However, beyond all those familiar reasons for imperfect adaptation, maladaptive behaviour may also be normal. It is a possible outcome of the principle of melioration itself, not only of something that has gone awry. A tennis player may be losing matches because he has equalized the average benefits of lobs and passing shots. He may be irrational in doing so, but not deranged. Less hypothetically, Alisdair Houston and Nicholas Davies have shown that the pied wagtail, foraging in the Oxford area, gathers less food than it could by allocating its time differently. It falls short, Houston suggests,[21] because the reinforcement contingencies happen to approximate one of those arrangements shown in the laboratory to produce suboptimal choices, by human subjects as well as animal.

Social Darwinism in Light of Psychology

The usual point of contact between psychology and social Darwinism is the study of heritable individual differences, such as intelligence and personality. People vary psychologically in socially significant ways, as everyone recognizes; the question is why. Controversy rages over whether or not the differences between people or groups of people have purely environmental origins. The balance of argument will be tipped in time by the weight of evidence, from differential psychology and from other disciplines that examine the nature of human variation—human biology, economics, sociology, and the rest of the social sciences. This is a quarrel about variation, one of the two pillars of Darwinian theory.

Reinforcement theory, in contrast, speaks primarily to the selection of behaviour, not its variation. It, too, bears on social Darwinism or sociobiology, and will in time stand or fall with the evidence. To critics of Darwinian thought who argue that behaviour is learned, rather than inherited, the answer from reinforcement theory is that experience and inheritance are inextricable. Behaviour is shaped by its consequences, showing the effects of environmental history. But what is effective as a reinforcer for behaviour has an inherited core. We may teach an organism to jump through hoops for food or to avoid pain, but the character of the reinforcing stimulus needed to influence behaviour is as much part of the organism's inheritance as the hoop is part of its environment. Higher organisms inherit a capacity to be reinforced by particular stimuli, which constitute a hereditary endowment no less rooted in biology than the instinctive actions of lower species.

The objects or events, or often the activities themselves, that reinforce the behaviour of a species comprise its psychological

[21] Houston, 'The Matching Law Applies to Wagtails' Foraging in the Wild'.

nature. Birds building nests, infants rising up on their legs to take their first steps or straining to articulate the sounds they hear, cattle congregating in a pasture, cats pouncing on mice, all these and countless other species-typical behaviours, are examples of reinforcers shaping behaviour while also expressing a genetic endowment. They are examples that show how the behaviour of individuals is cut and fitted by reinforcement for the environments in which evolutionary selection operated for the species.

Behaviour is malleable to the extent that it is feasible to devise suitable contingencies of reinforcement; this is a limitation that falls far short of perfect malleability. As a practical matter, it may be virtually impossible to train cats to graze in pastures and cows to pounce on mice, because those creatures do not respond to the appropriate reinforcers for the behaviours involved. Human behaviour may be primarily learned, as social reformers claim, but the limitations inherent in what reinforces us severely constrain the shapes human society can take. Reinforcement theory is, in this sense, the empirical realization of the ancient doctrine of human nature. Social reformers must come to terms, not only with heritable variation, but with heritable constraints on the selection of behaviour.

On the other hand, to those whose trust in the selective powers of evolution is such as to lead to *laissez-faire* conservatism or more extreme brands of right-wing social Darwinism, the answer is that we now know that natural selection cannot be trusted to do the right thing. The dynamics of evolution are different from genuine teleology—they do not necessarily find optimal solutions to the problems facing a species. This is also the answer to the ontogenetic version of social Darwinism, the libertarian or rationalist view that individuals or groups, behaving rationally, will maximize something worth maximizing. They will, in fact, not necessarily maximize social welfare—they may overfish or stand at football matches. They may not even maximize their own reinforcements—by meliorating, they may lose instead of gain. The invisible hand may guide us, but it may have mischief up its sleeve.

To Darwinians or their critics of the left or the right, the message from reinforcement theory is that to understand or control behaviour, we must think small, in the local terms of specific contingencies of reinforcement rather than in the global terms of general welfare or political philosophy. The United States has recently been alerted to the failures of its welfare policies, as in Charles Murray's much discussed book, *Losing Ground*.[22] Payments and subsidies to the needy, he suggests, have increased unemployment; aid to impoverished .single-parent families seems to have increased the rate of illegitimate births. Murray's analysis is a recent contribution to the literature of unintended consequences— social policies that exacerbate the very conditions they were meant to reform. By trying to lessen the burdens of unemployment or illegitimate parenthood, American welfare policy weakened the natural disincentives against them and the result was an increase in both.

Murray's examples, and many others, illustrate failures to have seen what the implicit reinforcement contingencies of a policy are. What behaviour does a welfare cheque or a tax schedule reinforce? If people were rational in the long term, they might use their resources constructively, but what if they are meliorators rather than maximizers? Reinforcers given to alleviate wants not only alleviate wants, they also strengthen the behaviours upon which they are conditional. In the long run, the strengthening may be more significant than the alleviation. Policies conceived in benevolence may instead reinforce indolence, bureaucratic manipulation, and profligacy. Given the reinforcers society has at its disposal, and the nature of the control exerted over behaviour by those reinforcers, how shall the contingencies of reinforcement be arranged? What are the latent dangers in establishing a particular control over the behaviour of citizens? Such are the proper questions for social policy. Nothing very general can be said about the answers, except that they are unlikely to fit comfortably into any familiar political persuasion of the left or right.

[22] *Losing Ground: American Social Policy, 1950–1980* (New York, 1984).

4

Incest Taboo and the Evolution of Society

MAURICE GODELIER

It is a great honour for me to be here with you today, and I would like to begin by expressing my deep gratitude to the organizers of the Herbert Spencer Lectures for having invited me to contribute to this continuing investigation into the evolution of life and society.

As we shall see in a few minutes' time, man possesses the society that his brain allows him to have. Not that the forms of society are pre-programmed in the brain, but in the sense that the generation of properly human social relationships always presupposes that they take shape initially inside and outside thought; and in the sense that these social relationships always imply on the part of individuals both individual and collective forms of understanding of what is happening inside themselves, among themselves, and through themselves. This is one way of interpreting Herbert Spencer's famous dictum that human society is a supraorganic phenomenon whose development is an extension of the biological development of the human species.

Some Features that Appear to be Specific to Man

To understand some of the mechanisms of the evolution of human society, I believe we first need to recall a number of essential facts that are specific to man.

Fact number one: humans, unlike other social animals, do not merely live in society, they create it. Humans produce society in order to live. That is why, not only do they have a history like other social animals, but they partly help to make their history. 'History' with a capital H. We shall see later that this history is not purely the work of a society's internal forces, but that it also springs from external forces, from chance encounters from the unexpected. In a word, human history is like the evolution of life, a mixture of chance and necessity.

Fact number two: humans, like all other living beings extract their means of existence from nature around them, but they do so by transforming this nature and by transforming their relationships with nature. By these transformations they modify not only their conditions of subsistence, but all the material conditions of their social existence, the material basis of society.[1]

The association of these two facts immediately raises the problem of whether or not there is some correspondence between the transformation and succession of forms of society on the one hand and the transformation and succession of forms of mankind's relationships with nature on the other. In the West, from the middle of the eighteenth century onwards, from Adam Smith or Quesnay to Morgan, Spencer, Tylor, via Marx and others, there was assumed to be some internal linkage between the demise of classless hunter-gatherer societies and the emergence of societies organized into hierarchies based on orders, castes, or classes, the name matters little. There have admittedly been exceptions, since societies as stratified as those of the North-west Coast or the Florida peninsula Indians in North America depended on an economy based primarily on fishing, hunting, and the gathering of wild food. But these exceptions, far from contradicting the foregoing examples, already point to one essential condition of the emergence of hereditary social hierarchies, namely the productivity of human labour which enables one part of society to live without directly participating in the material tasks of production; this leaves the people concerned free to devote their time to other functions that society holds to be essential. In the case of agricultural or specialized animal husbandry, this productivity is achieved through the exploitation of animal and vegetal species that have been domesticated by man. For the Kwakiutl or Calusa hunter-fisher folk, this was made possible by the exceptional

---

[1] Maurice Godelier, 'Lewis Henry Morgan', in *Encyclopedia Universalis* (Paris, 1971), xi. 323–5; id., 'Anthropology and Biology: Towards a New Form of Cooperation', *International Social Science Journal*, 26 (1974), 611–35; id., 'Infrastructures, Societies and History', *Current Anthropology*, 19 (1978), 763–71; Avant-propos to *L'Idéèl et le matériel* (Paris, 1984), Eng. trans., *The Mental and the Material* (London and New York, 1986).

abundance of wild resources capable of being exploited by them in their natural environment.

This too raises several questions, one of which is how and why changes occur in the intellectual and material capacities of human beings to act upon their surrounding ecosystems, because these capabilities plainly do not develop of their own accord. If they do develop they do so in specific social and historic circumstances and we want to know the reasons for and the workings of those circumstances. Another question is: would a change in our scale of investigation—a less exalted, less global look at historical evolution—reveal finer correspondences between the existence of a given system of kinship, a matrilineal one for instance, and a particular form of agriculture or animal husbandry. On this level, results to date have been disappointing, no anthropologist having yet succeeded in showing reasons why a particular type of economy necessarily coexists with a particular type of kinship system within a given society.

From these premises, I want to try to analyse what it means for man to produce society, by looking at relations of kinship.

Humanity, it needs to be remembered, *did not invent* society. It is the evolution of life which introduced into the animal world forms of existence that connect the development of individuals of different sexes and generations within a single whole, which reproduces itself and them. Nor did humanity *invent the family*, for one encounters this kind of social group in certain animal societies; what it has done is to add to the animal family relationships that have modified it radically and interposed relations of kinship between individuals, their families, and the community, all of which go to make up their society. These relationships are the product of principles and rules which explicitly determine the social position of an individual relative to all or some of his/her ascendants, and the conditions of his/her sexual intercourse and social alliance with an individual of the opposite sex. The existence of relations of kinship dividing society into groupings that run through and transcend family units is today a fact specific to *Homo sapiens sapiens* but may also have characterized certain of his closest ancestors.

Now the *Homo sapiens sapiens* that we are possesses the most complex nervous system and brain known to us. Our brain is characterized by the ability not only to grasp or construct relationships between things or individuals in a given context but also to grasp or construct relations of equivalence or of non-equivalence between these relationships, in other words, our brain is capable of perceiving or constructing *relationships between relationships*. Starting from this property of the brain, we can understand two other peculiarities of man.

It was long thought that what distinguished man from the higher primates was the fact of using tools, but over the last twenty years a considerable body of work has shown that man shared this faculty with a number of other species. It is now clear that man's distinctive difference lies in his ability, not to use tools, *but to manufacture tools in order to make other tools*.[2]

A second peculiarity of man is the existence of *articulated* language, of a means of communication to a large extent unconnected with the immediate bodily and environmental context of the speaker and capable of communicating and trans-mitting abstract ideas, i.e. representations of realities internal or external to man, tangible or otherwise, which can be perceived equally in terms of their relationships to the speaker, in terms of their reciprocal relationships, and in terms of their transforma-tions. It is because human language is subject to the workings of abstract thought that it has arisen through the use of a perceptible abstract material, namely the sounds of articulated language, i.e. the sounds of the different so-called natural languages. For the existence of an articulated language presupposes the existence of thought capable of perceiving the relationships between different realities within a particular context, and of comparing them with similar or contrasting contexts which no longer—or do not yet—exist.[3] In other words, it presupposes a form of thought capable of

---

[2] B. D. Beck, *Animal Tool Behavior: The Use and Manufacture of Tools by Animals* (New York, 1980); Tim Ingold, 'Notes on the Foraging Mode of Production', Paper delivered at the Fourth International Conference on Hunting and Gathering Societies (London, 1986); William C. McGrew, 'Evolutionary Implications of Sex Differences in Chimpanzee Predation and Tool Use', pp. 441–63 in D. A. Hamburg and E. R. McCown (eds.), *The Great Apes* (Menlow Park, Calif., 1979); McGrew, C. E. G. Tutin, and P. J. Baldwin, 'Chimpanzees, Tools and Termites: Cross Cultural Comparisons of Senegal, Tanzania and Rio Muni', *Man*, 14 (1979), 185–213; G. M. Guilmet, 'The Evolution of Tool-Using and Tool-Making Behavior', *Man*, 12 (1977), 33–47.

[3] P. Lieberman, *The Biology and Evolution of Language* (Cambridge, Mass., 1984); J. T. Laitamn, 'L'Origine du langage articulé', *La Recherche*, 17 (1986), 1164–86.

symbolically representing—in the form of concepts or images—a series of objects and relations perceived in their series of past, present, and future contexts. Not that thought can be reduced to language; but language is thought directed towards others and acting within a given society upon society. Abstract symbolic thought like articulated language therefore presupposes a mastery of time and space in such a way as to allow a past that no longer exists and a future that does not yet exist to coexist in the present. These two peculiarities have been, and will remain, the essential preconditions of the production of relations of kinship.

Origin and Foundations of the Prohibition of Incest

The first thing to note is that in all societies we find one fundamental prohibition which is expressed in various forms and appears to be the prime condition of the existence of relations of kinship, namely the prohibition of incest, or the ban on hetero-sexual relations between the members of the same family, and more broadly between certain categories of consanguines. How are we to account for the universality of this taboo in the most varied societies and at all those periods of history for which we have evidence? Because the taboo is universal, we cannot seek its causes in the particular content of each of these cultures. To account for it, we must look to the kinds of phenomena that have affected, and continue to affect, the very existence of all human societies.

I shall now attempt an explanatory model of the conditions in which the incest taboo may have emerged in the course of the transformation of our hominid ancestors first into *homo sapiens* and subsequently into *homo sapiens sapiens*.

First though, a brief reminder of some previously attempted explanations. As Françoise Héritier-Augé has pointed out in an important article, all these explanations ultimately come down either to efficient causes connected with nature, or to final causes connected with culture, with society.[4]

---

[4] F. Héritier-Augé, 'Incesto', *Enciclopedia Einaudi* (1979), vii. 234–62; Norbert Bischof, 'The Biological Foundations of the Incest Taboo', *Social Sciences Information* (1975), 7–36; id., 'Comparative Ethology of Incest Avoidance', in R. Fox, *Biosocial*

Where nature is concerned, some biologists have suggested that mechanisms analogous to the incest taboo are observable in certain species of primates close to man, and that these mechanisms have come to full fruition in man. Today, students of baboon behaviour, such as Kummer, or of chimpanzee behaviour, such as Kawanaka or Itani, cautiously advance the view that there may be biological mechanisms in these primates that prevent sexual relations between a mother and her young adult sons.[5] From my reading of their work and from my discussions with Kummer I think we are dealing here not with a practice corresponding to the prohibition of incest, but rather with a consequence of the mechanisms of *the relationships of hierarchy and domination* that organize bands of primates, which creates a situation resembling the effects of an incest taboo.[6]

Because of the existence of hierarchies among males, and between males and females whose relations occur within a climate of sexual competition, a young male would find it socially impossible to gain sexual access to his mother. But no specialist would claim that the prime purpose of a social hierarchy and of relations of domination among primates was to serve as a biological mechanism to prevent and eliminate sexual relations between blood relations. This hierarchy performs other well-known functions, such as to ensure the protection of females and the young, and more generally to protect the band from danger

---

Anthropology (New York, 1975), pp. 37–67; Leslie-A. White, 'The Definition and Prohibition of Incest', *American Anthropologist*, 50 (1948), 416–35.

[5] J. Itani, 'The Evolution of Primate Social Structures', *Man* 20 (1985), 593–611; J. Else, and P. Lee, *Primate Ontogeny, Cognition and Social Behavior* (Cambridge, 1986).

[6] H. Kummer, *Primate Society: Group Techniques of Sociological Adaptation* (New York, 1971). This view was in one sense already put forward by Darwin in *The Descent of Man*, ii, chap. 20: 'According to what we know about jealousy in all mammals, many of which are armed with special organs to help them in their struggle with rivals, we may conclude that *general promiscuity* of the sexes in a state of nature is *extremely unlikely*.' (emphasis added.) It is interesting to note, moreover, that although Freud mentions Darwin in *Totem and Taboo* and stresses the originality of Darwin's thesis, he then does nothing about it, other than to comment that 'in practice, the conditions which Darwin assigns to the primitive band could foster only exogamy'. He then drops the question of the origins of incest, arguing that it is one possible, though partial, explanation of the problem of 'exogamy': 'I must also mention one last attempted explanation of incest. This attempt is *totally different* from those that we have dealt with so far and might be described as *historical*. It flows from Charles Darwin's hypothesis concerning the primitive social state of Humanity. From the habits of the higher primates, Darwin concluded that man too had originally lived in small bands, within which the jealousy of the oldest and strongest male would prevent sexual promiscuity.' (p. 145, emphasis added.)

and attack by predators. Before leaving this argument, however, there is one feature of it that could help us later on to elucidate an aspect of human relations of kinship, namely the existence in certain primates of a social hierarchy, and relative domination by the males in contexts of confrontation and protection.

Of course mother–son incest is not the only form of incest: we also encounter father–daughter incest and, above all because this affects the future through the younger generations, incest between brothers and sisters. We need a good deal more empirical observation to find out which of these incestuous relationships is systematically eliminated by hierarchic relationships in primates. But one can raise one general objection on a point of principle to all these empirical studies. This objection was formulated at the turn of the century by Frazer in his writings on Totemism and Exogamy:

It is hard to see why a deep-rooted human instinct should need to be reinforced by a law. There is no law telling man to eat and drink, or forbidding him to put his hands in the fire. What nature herself prohibits and punishes has no need to be forbidden and punished by law. Thus, instead of inferring from the legal prohibition of incest that there must be a natural aversion from it we ought rather to conclude that there is a natural instinct that drives people to commit incest and that if the law reproves this instinct like so many other natural instincts, it is because civilized man has realized that the satisfaction of these natural instincts would be harmful from the point of view of social development.[7]

Freud mentions Frazer's argument in 'Totem and Taboo', showing how it ties in with the conclusions of psychoanalytical experience, which reveals the spontaneous tendencies to incestuous desire between members of the nuclear family.[8] This suggests

---

[7] J. G. Frazer, 'Totemism and Exogamy' (1910), iv. 97. Frazer levels this criticism at Westermarck who, in his work on the origin and development of moral concepts, accounted for the fear of incest as follows: 'People of different sexes, living together from their earliest infancy feel an innate aversion to engaging in sexual relations and, since there are generally blood ties between these people, this feeling finds in custom and law its natural expression, namely the prohibition of sexual relations between close relatives.' *Ursprung und Entwicklung der Moralbegriffe*, vol. ii (1909).

[8] S. Freud, *Totem et Tabou*, trans. S. Jankelevitch (Paris, 1965), p. 143. Freud sides with Frazer in his criticism of Westermarck: 'Where, ultimately, does the fear of incest, which we should regard as the very root of exogamy, arise from? It is obviously not enough to account for the incest phobia by an instinctive aversion to sexual relations between very close relatives, which is tantamount to putting it down to the fear itself of incest, whereas

rather that if human sexuality were left in its 'savage state', it would fasten on any possible partner, with no regard to what is 'appropriate' in a given society, still less to what is obligatory.[9] The thousands of cases of incest reported to the police each year in the West are evidence that even today, in our own societies, the power of prohibition on incest remains limited, notwithstanding moral opprobrium, the threats of prosecution and punishment, or religious condemnation.

I shall leave to one side for the time being the question as to whether the evolution of nature created in man's immediate ancestors a situation that rendered it essential for society to regulate sexual relations through a rule binding on everyone and on all generations. I want to turn to the other type of explanation, the teleological view, according to which human beings acted deliberately to bring order to society by inventing relations of kinship. The outstanding exponent of this view is, of course, Claude Lévi-Strauss, who places it at the heart of his 'Elementary Structures of Kinship'. He summarizes this view in an article entitled 'The Family', published in 1956:

As Tylor showed almost a century ago, the last explanation, probably, is that humanity grasped very early on that, to break free of a savage struggle for existence, it faced a stark choice: either marrying out or being killed out. It had to choose between isolated biological families juxtaposed like discrete units, perpetuating themselves by their own means, submerged by their dreads, their hatreds and their ignorance and, thanks to the prohibition of incest, the systematic institution of chains of intermarriage, making it possible to build an authentic human society on the artificial basis of bonds of affinity, notwithstanding, and even in opposition to, the isolating influence of consanguinity.[10]

It is worth dwelling a few moments on this quotation and the picture of primitive life that it evokes. It was this vision of savage

experience shows us that, notwithstanding this instinct, incest is far from uncommon, and where historical experience teaches us that incestuous marriages were obligatory for certain privileged persons.' (ibid. 141).

[9] Maurice Godelier, 'Le Sexe comme fondement ultime de l'ordre social et cosmique chez les Baruya de Nouvelle-Guinée', in A. Verdiglione (ed.), *Sexualité et pouvoir* (Paris, 1976), 268–306.

[10] C. Lévi-Strauss, 'The Family', in H. Shapiro and G. Dole (eds.), *Man, Culture, and Society* (Oxford, 1956), pp. 261–85; E. B. Tylor, 'On a Method of Investigating the Development of Institutions Applied to Laws of Marriage and Descent', *Journal of the Anthropological Institute*, 18 (1888), 245–69 (p. 267).

struggle, of dreads, hatreds, and ignorance which, he argues, ultimately drove humanity to create a new social order, by binding each individual with ties of solidarity born out of the renouncement of incest and the institution of exogamy and the exchange of women. This was the triumph of reason over primitive, brutal, selfish sexuality, the substitution of reciprocity in the service of enlightened self-interest, and in the interest of all, for the egotism, hatred, and fear of primitive animal families.

Note in passing that this quotation contains two presuppositions on the part of Lévi-Strauss, namely (i) that consanguinity is a natural tie, which for him implies that the biological, animal family of our ancestors must have recognized *both* paternity and maternity; and (ii) that among these animal families there can have been *no foundation* for co-operation, sharing, and reciprocity *apart from* sexuality. In other words, to build an 'authentic' (?) human society in which solidarity, co-operation, and the defence of the common good play a part, man had to, and did so with success, invent artificial relationships of matrimonial alliance, adding them to the more natural animal consanguinity. Hence the series of apparently rigorously transitive implications which lie at the basis of Lévi-Strauss's theory: incest entails exogamy, marriage outside one's people, and the exchange of women, because, according to Lévi-Strauss, by shunning 'marriage with mother, sister or daughter', each man bows to a 'rule which obliges him to give mother, sister or daughter to another man'.[11]

Before taking issue with this explanation of the origins of incest and the syllogism that it seems necessarily to entail, I should say straight away that it represents a fundamental theoretical step forward, and that there can be no question of going back on that. The prohibition of incest is an *adaptive* response on the part of humanity to practices connected with sexuality which could jeopardize the very survival of society, i.e. the necessary ties and dependence that bind humans together, and to society, band, or human community to which they belong, in their existence and in their development: humanity 'has grasped the point'.

Other criticisms than the ones I am about to spell out have been levelled at Lévi-Strauss. For example, authors such as Moscovici

[11] C. Lévi-Strauss, *Les Structures élémentaires de la parenté* (2nd edn., Paris, 1967).

or Peter J. Wilson claim that the invention of relations of kinship is not primarily the invention of relations of alliance, but first and foremost the invention of social paternity. In primate societies, in which we can observe patterns of recognition of their young by mothers throughout their life span, we apparently still lack conclusive evidence that the fathers continue to recognize their young once the latter have left their family of origin. It is therefore argued that relations of kinship arose with the invention of the Father.[12]

For Peter Wilson, though, this invention did not spring from any human decision, but rather from the accidental encounter of two biological relationships which are also to be found in primates, namely the enduring, continuing, intense, reciprocal primary bond between an adult female and her descendants, and the bond between a male and his adult female or females. These two bonds are thought to have undergone a series of essential modifications in the course of the evolution of the human primate, which would have, as it were, 'produced' the father.

On the one hand the fact that human young mature much later would prolong and intensify the mother–child relationship. On the other, by making them sexually receptive all year round, the loss of oestrus in human females would create a much more intense and more enduring bond of attraction between the male and his adult female or females, this being the first step towards the emergence of social paternity. The necessary conjunction of these two relationships having acquired a fresh dimension due to the process of biological evolution, would thus for the first time create the conditions for the emergence of specifically human social relationships, that is of a relationship between two social relationships, or to put it another way, for the emergence of a meta-relationship.

From being the temporary protector of females and the young, caught up in relations of domination or subordination *vis-à-vis* other adult males, the male, it is argued, was transformed into a father by the extension and the intensification of his bonds with the females and the young. Further, the conjunction of the two

[12] Peter J. Wilson, 'The Promising Primate', *Man*, 10 (1975), 5–20; id., 'La Pensée alimentaire: The Evolutionary Context of Objective Thought', *Man*, 12 (1977), 320–35; Serge Moscovici, *Essai sur l'histoire humaine de la nature* (Paris, 1968).

relationships is supposed to have created a situation in which our ancestors, possessing a brain comparable to our own, came to *make decisions* as between what they accorded themselves (between males and females) in the framework of increasingly durable sexual ties, and what they accorded to their descendants. In other words, human kinship is thought to have arisen with the emergence of social paternity. The paradox of this theory is that, on this view, the incest taboo is not at all a condition of the production of relations of kinship; it is merely a condition of their reproduction. As for alliance and marriage, according to the author these become 'the formalization of the institution of relations between *males among themselves* and through the intermediary of females'. The relations thus instituted complete the emergence of the relations of kinship.

This theory accounts for the emergence of kinship without referring to the incest taboo, and it presents this emergence as the necessary outcome of the 'accidental and external' conjunction of two biological relationships, with final causes rounding off the process. For, according to the author, the transformation of the two relationships created a situation in which, for the first time in the evolution of animal species, individuals were able and obliged, in their sexual relations, to refer and to defer to their relationships with their descendants, and vice versa. And so, with this 'materialization of the future' and the taking into account of supra-personal relationships, we are brought before the 'very seeds of society'.

I now want to try to propose an alternative interpretation of these facts. The loss of oestrus in the human female (for which loss biologists have yet to supply a satisfactory explanation) does indeed entail the permanent possibility for all the adults in a band or society, male and female, young and old, to engage in generalized sexual exchange. Meanwhile, the prolonged period of maturation in children entails the presence within a more permanent animal family of descendants, some of whom at least have attained puberty and can also therefore enter the generalized sexual fray.

Now let us compare the situation created by the *accumulation* of these two biological transformations with two other elements pertaining to the social and material reproduction of our primate ancestors, namely on the one hand the existence of sex and generation-determined relations of hierarchy, dominance, and subordination: this is true of both the sexual sphere and the quest for subsistence. On the other hand, thanks to the possibilities opened up by the development of the brain, bipedal locomotion, and the increasingly intensive utilization of implements, these individuals are able to *co-operate* in ways more complex than non-human primates in the material and social reproduction of their existence. And this, needless to say, in a context of strategies regarding the exploitation of spontaneously available resources, abundant or scarce, but always uncertain, in the natural environments in which our ancestors found themselves competing with other predatory species.

This comparison suggests that an objective *contradiction* has arisen and developed between these four factors, especially between the broadening of the sexual sphere, now unshackled from the rhythms of nature and short mating seasons, and the broadening of the sphere of material and social co-operation. For this, of course, we need to consider that the broadening of the sexual sphere has consequently intensified the competition and hierarchic relationships that are also characteristic of these societies. There thus arose, wherever this stage of biological and social development had been reached a situation that demanded the *conscious deliberate intervention* of men to *control and regulate sexuality* in such a way that it would once more be made *subordinate* to the reproduction of society.

This interpretation, it should be stressed, provides a possible explanation of how the prohibition of incest came into being, but not a word has been said yet about relations of kinship. We are thus in a position that is both the reverse of Wilson's position, since for him there is no accounting for the emergence of kinship with the prohibition of incest, and distinct from that of Lévi-Strauss, for whom the purpose of this prohibition was to institute relations of kinship by adding the artificial dimension of relations of alliance to the quasi-natural dimension of consanguinity. For with the prohibition of incest there emerges a phenomenon of general social significance, which encapsulates as it were all the characteristic features of human *society*. And it is only at another, more restricted, level that the prohibition of incest functions as the necessary pre-condition of the emergence of relations of kinship.

These should therefore been seen, not as the goal pursued by the institution of this sexual taboo, but as its quasi-automatic, and indeed almost involuntary, outcome. The reason for the emergence of relations of kinship as soon as sexual relations within the animal family had been forbidden should be obvious, namely that, if it is forbidden, impermissible for the members of animal families to have sexual relations with each other even though they are physically capable of doing so, then they are obliged to have those relations outside the family. If everyone observes this prohibition then they are automatically, as it were, obliged to engage in a social exchange of sexual partners between different families. This was what Lévi-Strauss meant when he wrote that the prohibition of incest is 'less a rule that forbids a man to marry his mother, sister or daughter, than a rule obliging him to give his mother, sister or daughter to someone else.' (cf. n. 11.)

So the fact of observing the incest taboo simultaneously gives rise to relations of kinship in all their dimensions. For when one rules out the possibility of choosing a mate from within one's own family and must therefore look elsewhere, and, when this rule becomes applicable down the generations, one needs to define, identify, and memorize just who one *is* and who one *is not*. One needs to recall the men and the women from whom one is descended. Once the obligation to exchange is applicable to all individuals and their families, each person is obliged to define his social identity and to preserve that identity by reproducing it from generation to generation. It is in this sense, and through this process, that the prohibition of incest entails—as a consequence, not as a goal—the *simultaneous* appearance of *both* axes of kinship, namely *filiation* and *alliance*.

But as soon as relations of kinship emerge, the prohibition of incest embarks on a *second* career. From having been a necessary pre-condition to the emergence and production of relations of kinship, it now becomes the permanent condition of their *reproduction*. It lodges in them and reproduces with them. However, as in its original emergence and motivation, the prohibition on incest was bound up with the survival of society as such, its effects immediately spill beyond the sphere of kinship. In one sense it encapsulates the whole of human society, which is why anthropologists and sociologists are not alone in their exploration of it: it intrigues psychoanalysts, and even poets such as Goethe, whose Wilhelm Meister turns on this theme just as much, albeit in other ways.

Relations of Kinship: Properties, Functions, Evolution

Let us now take a closer look at the general properties of relations of kinship. Kinship is not just recognition of father, mother, father-in-law or mother-in-law. But it is equally and just as much knowledge of the relationships of father's father, father's father's father, mother's mother, mother's mother's mother, and so on. This, then, entails recognition of a *network* of *abstract, transitive* relationships, which in turn presupposes the ability to perceive relations of *equivalence* or non-equivalence between these relationships. For example, the equivalence of the relationship between a son and his father, and of the relationship between the father and his own father. But at the same time, it implies the realization that, in this relationship, the son–father relationship is not equivalent to the father–son relationship.

All these different relationships taken together would appear to be indefinitely extensible in principle; but in all systems of kinship, *only* a greater or smaller sub-system of these relationships, centred around an abstract ego, *finds expression in language*, in the form of descriptive or classificatory terminologies of kinship. The fact that all kinship terminologies refer to an abstract ego, characterized solely by the fact of being either a man or a woman, is evidence that here the individual is conceived purely as the point of arrival and departure of a set of abstract relationships, which then reproduce through him and by him.

But relations of kinship define not merely the identity of an individual in relation to other individuals (all those with whom he enters into relations of descent or alliance). Also, on the basis of all or some of these relations of descent, they define *social groups* of kinship, moieties, lineages, clans, kindreds, etc., to which each individual belongs, generally from birth, and among which all the members of society are distributed. These kinship groups enter into alliances when their individual members do so. Far more so than the individuals, the groups are the key partners in matrimonial exchanges and in all the other forms of exchange that marriage entails prior to and after celebration, including gifts of game, exchanges of labour, solidarity in feuds, and all the different kinds of material gifts and services that affines are supposed to give each other; these are continued in subsequent generations in the form of the gifts and services that one owes to or receives from one's maternal kin.

But the immediate outcome of the alliance between individuals (and through them the alliance between kinship groups) is not the production of other individuals: it is the production of a family which, like the individual, is thus defined from the outset by the structure of the relations of kinship that produce it and which it reproduces. In this way the animal family, an institution inherited from nature, preserves its original functions as the locus of sexual relations between adult males and females and of the production and rearing of children; but it is now also pressed into the service of and subordinated to the reproduction of a broader range of social relations, which permeate the family through and through, reshaping both its structures and its functions.

For the first time in the evolution of animal societies, the advent of relations of kinship as a means of controlling and regulating sexuality interposes, between the individual and the animal family on the one hand, and the band or society on the other, social groups of a kind never before seen in nature. These groups are founded on a web of abstract relationships which constitute a specifically *human* form of division of society.

Two remarks are called for here: it should be clear now why I opened my lecture with the brusque comment that men have the society that their brain permits. It is now plain that the emergence of abstract, transitive, etc. relations of kinship presupposes a thought process capable of perceiving both relationships, and relationships between relationships, as well as being capable of *decontextualizing* these relationships by understanding their conditions of reproduction, as witnessed by the fact that all systems of kinship refer to the existence of an unspecified, abstract ego.

In this sense my belief is that there is no such thing, and there

could be no such thing, as relations of *kinship* among animals. They do have groupings which structure the animal family, around either the male or the female, thus producing a simulacrum of patrilineal or matrilineal relations of kinship.[13] But, for relations of kinship to exist, these relations need to be *recognized and understood* and they must be regarded as the *principle* underlying specific forms of social behaviour as a framework for obligations and rights. Seen thus, it is clearly pointless to claim that kinship is primarily a question of inventing the Father. Kinship is neither just the father nor just the mother, nor the father-in-law nor the mother-in-law: it implies the immediate and necessary emergence of relations that posit and impose the existence and recognition of the father's father, of the mother's mother, the father's sister, the mother's brother, and so on.

It is rather easier now to understand why the institution of relations of kinship *does not entail* even approximate knowledge of reproductive biology, or of the role of men and women in this process. The direct effect of relations of kinship, through regulation of sexuality, is to *attribute* individuals, the children born from alliances, to one or another of the groups involved in the exchange. So all that emerges from the workings of relations of kinship is the individual's *social*, not biological, identity. It is by virtue of this social identity that each individual belongs, from birth, to one or another or to several of the kinship groups making up society, which reproduce in him and through him.

To give two examples of the relationship between the structures of a system of kinship and the ways in which societies view reproductive processes, I shall refer to the Baruya and to the Trobriand islanders. The Baruya live in the interior of New Guinea; they are patrilineal, and for them sperm alone makes children;[14] for the Trobriand islanders, who are matrilineal, the child is the product of a mingling of a spirit of a spiritual essence belonging to the mother's clan, with female bodily substances such as menstrual blood, for instance. Sperm is regarded as food for the foetus, and as something that gives it its shape. But in no sense is it

[13] R. Fox, *Biosocial Anthropology*; V. Reynolds, 'Kinship and the Family in Monkeys, Apes and Man', *Man*, 3 (1968), 209–23.

[14] Maurice Godelier, *The Making of Great Men* (Cambridge, 1986), chap. 3: The Institution and Legitimization of Male Superiority.

the direct agent of the conception of the infant.[15] One can see from these examples how ideas concerning the reproduction of life are structured by the nature of relations of kinship, which lay down the rules governing an individual's membership of a group of filiation which is much vaster than himself. Among the Baruya, children belong to the father's lineage: among the Trobrianders, they belong to the mother's clan. It is this sphere, in which the structures of systems of kinship articulate with systems of representation of the individual, the body, etc., that Françoise Héritier-Augé deals with in 'La Symbolique de l'inceste'.[16]

This shows why, although filiation and alliance are two distinct though complementary sets of relationships inside the system of kinship, filiation does appear by its functions to prevail over alliance. One must bear in mind that there are only four logical possibilities for determining an individual's filiation. Either the child is descended through the father, as is the case with patrilineal systems; through the mother, as in matrilineal systems; through father and mother, as with bilineal systems; or through father, mother, and their ascendants, as with cognatic systems.

Unfortunately, we still lack the means to understand the reasons leading a society to select one principle rather than another on which to build its relations of kinship. All we know is that all pastoral societies are patrilineal, with the exception of the Touareg, whose system contains a matrilineal element. In fact, one finds practically each of these principles of descent associated with widely differing patterns of subsistence, and the social sciences have not yet succeeded in analysing the processes whereby a given mode of production coexists with a given mode of filiation within a particular society.

The *abstract* nature of relations of kinship, and the fact they are organized into *networks* that are both centred on the individual and *off-centre* relative to him, enable them to serve as a *substrate for a whole series of concrete functions*. They may provide a framework for the organization of hunting or gathering, or more

---

[15] Annette Weiner, *Women of Value, Men of Renown: New Perspectives in Trobriand Exchange* (Austin, 1976); ead., 'The Reproductive Model in Trobriand Society', *Man*, 11 (1978), 175–86.

[16] F. Héritier-Augé, 'Symbolique de l'inceste et de sa prohibition', in M. Izard and P. Smith (eds.), *La Fonction symbolique* (Paris, 1979), 209–43.

generally for the appropriation of nature. They may also serve as a framework for the organization of relations between men and the powers that reign over society and the surrounding nature; they serve as a framework for religious and ritual practices. So relations of kinship can serve as the framework for a wide variety of forms of co-operation and sharing.

But all these networks of solidarity, co-operation, and sharing have their *source* and their *strength* in the crucial fact that kinship exists only because each individual creates for the others the conditions of their social existence and preservation, while receiving from them the conditions of his own social existence and the continuity of his own group. By degrees then, relations of kinship create a situation of *general* reciprocal dependence among all the individuals and groups in a given society; this dependence is inaugurated by the exchange of live individuals between groups, in which the act of transfer is simultaneously an act of *production* and of reproduction of social *relationships*.

Man therefore is apparently the only animal species to exchange *live* individuals between groups and their constituent individuals; in fact this physical exchange involves both tangible and intangible social phenomena, producing and reproducing relationships. The problem then is to know which individuals are exchanged. The usual answer is the familiar one, which is central to Lévi-Strauss's theory, namely that it is women, not men, who are exchanged; and that it is men who exchange women among themselves in their capacity as representatives of their kinship group.

However, a closer look at Lévi-Strauss's arguments quickly shows that he fails to supply any theoretical proof in support of his conclusion, for what does he in fact demonstrate? The existence of a logical sequence between three operations, which are presented as three interdependent conditions of the production of relations of kinship. The moment the prohibition of incest (moment number one) is established, it forces (moment number two) men and women alike to seek their spouse outside—i.e. exogamy—and obliges (moment number three) each to yield or give in exchange the sexual partners he has renounced. Thus, there is only one conclusion to be drawn from Lévi-Strauss's reasoning, namely, that kinship is exchange. Yet, when summing up the essence of relations of kinship, he employs two formulas, which he presents as equivalent: kinship is exchange, or, kinship is exchange of women.

These formulas are not equivalent. The first is a general one, applicable in all circumstances, presupposing no particular relation of power between the sexes in order to make relations of kinship work. The second, on the other, is particular in scope and is logically vitiated from a theoretical standpoint. For to deduce from the prohibition of incest that kinship equals the exchange of women, Lévi-Strauss would have to introduce into his theoretical thinking a fact for which no explanation is given in his theory, namely, the general domination of men over women in human societies.

Now in terms of social logic the prohibition of incest simultaneously opens up three logical possibilities of exchange: that men exchange women among themselves, which assumes that they dominate women in society, that women may exchange men among themselves, which assumes that it is they that dominate men; or that groups may exchange men and women among themselves, which implies no domination of one sex by the other a priori.

So Lévi-Strauss has introduced into his theory as a matter of course a fact for which he produces no theoretical explanation. In his writings, consequently, male domination turns out to have two different, and ultimately complementary, meanings. On the one hand, it is presented as a natural phenomenon which underlines our resemblance to the other species of primates in which males dominate females, and as an insuperable legacy of the biological evolution of our species. On the other, it is presented as the *sole* logical course open to human beings, which is manifestly not so.

Of course, Lévi-Strauss is well aware that there are three logical possibilities regarding the exchange of spouses, but, significantly, when he mentions the other two possibilities, he rejects them in mocking terms, as possibilities that can be raised purely to console ourselves for the harsh realities of male domination. In other words, he dismisses them as illusions that people, and especially women, may care to entertain about themselves:

Let my female readers, who may be shocked to see women treated as common chattels, as articles to be used in transactions between male operators, be assured *for their consolation* that the rules would be the same were we to choose to regard men as objects of exchange between female groups. Indeed, a few very extreme matrilineal-type societies have, up to a point, tried to *express* things in this manner. And both sexes may draw *satisfaction* from a different slightly more complicated way of describing the same process, which amounts to saying that groups of blood relatives, consisting of both men and women, undertake to exchange ties of kinship among themselves.[17]

In fact Lévi-Strauss goes so far that, having poured scorn on the suggestion of female domination as well as on the idea of equality of the sexes, he definitively announces women's inferior social status to be attributable to the properties of our brain. His conclusion to 'Elementary structures of kinship' demonstrates just how far apart we stand in our respective analyses of symbolic thought. There he states:

The emergence of symbolic thought *had* to require that women, like words, be things to be exchanged. In this new case, this was indeed the sole means of overcoming the contradiction which presented the same form in two incompatible aspects: on the one hand, as an object of desire, hence capable of exciting sexual and appropriative instincts; at the same time, as the subject, and perceived to be such, of another person's desire, in other words a means of binding it by binding her to oneself. But woman could not be a sign and nothing more, since even in a man's world she remains a person, and in so far as one defines her as a sign one is bound to recognize her as a producer of signs.[18]

Without dwelling on these passages, which speak for themselves, I would simply point out that the rules could never have been the same, unless one is prepared to say that in all three cases there would have been *exchanges of live individuals* between groups if women had dominated society, or if the sexes had been equal in society. The principle of exchange would have been the same in formal terms but the content of social life and the organization of society would have been different. It is a little surprising, therefore, to see feminists such as Gayle Rubin, who praises Lévi-Strauss for having revealed the oppression of women inside relations of kinship, unable to view Lévi-Strauss's ideo-

---

[17] Lévi-Strauss, 'The Family', p. 356.
[18] Id., *Les Structures élémentaires*, p. 569.

logical presuppositions critically and thus forced either to abandon their aspiration to liberate women, or else to forge ahead regardless of the catastrophic consequences that that could entail for humanity.

The 'exchange of women' is a seductive and powerful concept. It is attractive in that it places the oppression of women within social systems, rather than in biology. Moreover, it suggests that we look for the ultimate locus of women's oppression within the traffic in women, rather than within the traffic in merchandise ... The 'exchange of women' is also a problematic concept. Since Lévi-Strauss argues that the incest taboo and the results of its application constitute the origin of culture, it can be deduced that the *world historical defeat of women occurred with the origin of culture and is a prerequisite of culture*. If his analysis is adopted in its pure form, the feminist program must include a task even more onerous than the extermination of men; it must attempt to get rid of culture and substitute some entirely new phenomena on the face of the earth.[19]

Let me make myself clear. I am not criticizing Lévi-Strauss for having considered male domination to be a quasi-universal phenomenon, nor for having assumed that the situation must have been more or less the same for our ancestors. What I am criticizing is the fact that he has treated women's social subordination as something that is ultimately part of our biological nature, and which cannot, therefore, be changed at any time in the social evolution of humanity, throughout its history in other words, through any of the various forms of social organization that arise out of changes in men's relationships between themselves and with nature.

As we have seen, kinship as such has no need of male domination over women in order to function. It is when this domination exists for other reasons that it permeates the mechanisms of relations of kinship, pressing them into the service of its own reproduction.

For a fuller account of the organization of early human societies, where different forms of domination between the sexes and generations arose although remaining casteless or classless

---

[19] Gayle Rubin, 'The Traffic in Women', in R. Reiter (ed.), *Towards an Anthropology of Women* (New York, 1975), pp. 175–6.

societies, we shall need not only to seek the reasons for such inequalities elsewhere than in kinship, but also to distinguish the relative contributions of violence and consent to the generation and reproduction of these forms of domination. We should, for example, strive to understand how and why the sexual division of labour in primitive societies places the instruments and techniques of big game hunting and war in the hands of the men, thereby giving them a monopoly of armed violence. Yet we may be sure that men invented neither this division nor the monopoly with the intention of imposing them upon women; in all probability, these must have been thought of and experienced as being to the advantage of everyone, men and women alike, and thus beneficial to society.

There is not the time to discuss these issues here, but it is important to stress that, from the point of view of the evolution of human society, the existence of relations of kinship based on various forms of exchange of women between men has provided a medium for the development of numerous forms of individual and collective bondage between the sexes and the generations. In some cases these have gone as far as exploitation of women and youth by adult men, with regard to their labour, to the control and distribution of resources, and to sexual relations. On the fringes of relations of kinship, when unfavourable circumstances sever the links between an individual and his group, enslavement may occur. But it is also worth recalling that in the hunter-gatherer societies, where differences in individuals' and groups' capacities of accumulation are small, and where accumulation of goods is frequently condemned and suppressed by the community, the human being himself is, together with the resources in nature, the condition of society's permanence. So one can see what a formidable source of power the control of women, and through them children, must have represented. For women generally had less say than men in the fate of their children.

To come anywhere near the complexities of real life any analysis of classless societies, whether ancient or contemporary, also needs to examine the foundations of the various forms of communal or individual appropriation of the resources exploited; of forms of ownership of implements and weapons, etc. To understand those features of human existence such as the use of fire, common property, etc., we should look elsewhere than at this pervasive human sexuality. We also need to explore the complex sphere of relations between men and the higher powers, the spirits, ancestors, and gods which are thought to control the workings of nature and society, and which are the focus of religious ideas and rituals.

For, just as various forms of individual or collective subordination among the members of society arise within relations of kinship, so men's relations with what we call the supernatural powers serve as a framework within which to learn and acknowledge the general subordination of humanity as a whole, of men and women alike, to the powers that reign over it. I would, therefore, hypothesize that it was within the element of religious ideas and practice that the earliest rudiments of classes may have been born, as certain kinship groups and individuals became full-time specialists in the performance of rituals to maintain communication between men and gods, to the benefit of the community, and as these groups or individuals came to be definitively removed from any kind of productive work.

But here we are dealing with other moments in and other forms of man's social evolution, with the formation of hierarchic relations of a new type, relations which we, with our westerners' vision, refer to as relations between orders, castes, or classes. These relations go beyond relations of kinship in dividing up society, subordinating the latter to themselves. I shall return to this in a moment, after having essayed a provisional assessment of what we are entitled to conclude from the foregoing discussion of relations of kinship, their genesis, and their place in the evolution of society.

The Place of Relations of Kinship in the Evolution of Nature and Society

Relations of kinship are a fact of human life that has no equivalent in any of the animal societies that have evolved on Earth. Simplifying considerably, we may say that relations of kinship spell out the social conditions in which an individual may marry and have children; they also spell out the social identity of children born of his/her alliances, and identify those groups and individuals who have rights and duties *vis-à-vis* his/her children. Relations of kinship necessarily presuppose social control and conscious regulation of both sexual relations and the production and upbringing of children.

These are their *specific* and *universal* functions to which particular societies and particular moments in history may add other social functions: for instance, they may serve as a framework for ritual activities or for various forms of organization of labour and appropriation of nature. Regardless of the number and variety of their functions, in all societies, relations of kinship also serve as vectors for the transmission of each group's rights to certain statuses, wealth, territory, powers, etc. It is worth pointing out, in passing, that when performing their specific functions of regulating questions of filiation and alliance, relations of kinship can in no way be said to function as the superstructure of other social relations.

All systems of kinship presuppose the prohibition of incest, whose meaning and origins it is important to understand. Now, I have tried to show that the prohibition of incest cannot have been introduced for the purpose of creating human kinship, and that the formation of relations of kinship between men was its necessary consequence, not its goal. Not that humanity had no consciously acknowledged purpose when prohibiting incest. Simply, that purpose was not to create kinship but to safeguard society, the mode of existence, and the development of our species, as it sought to contend with the dangers and disorders associated with the emergence of a sexuality which the rhythms of nature had ceased to hold in check. The prohibition of incest is therefore a human institution which, by its origins and its functions, is not confined to matters of kinship alone. It has its origins in the evolution of nature, but it remains a continuously originating condition of the reproduction of human society. It is a negative expression of the law of human development which states that society can only exist by dividing while at the same time overcoming the resulting divisions.

So our analysis of the prohibition of incest reveals the existence of both a continuity and a discontinuity between the evolution of nature and the evolution of human society, between nature and culture. I started out from the fact that man did not invent society, any more than he invented the animal family. What he did do was to produce hitherto non-existent social conditions for the biological reproduction of life. I believe that he did so under the twin pressures of efficient causes which sprang from the evolution of nature, and of final causes which flowed from his own realization of the need to control sexuality in order to preserve his social mode of existence.

We have put forward the hypothesis that, among these efficient causes, two could have directly brought about a *social* situation capable of threatening the reproduction of society, namely the loss of oestrus in human females combined with the slow maturation of the human young. These biological transformations were in no way intentional. They occurred in a particular species of primates in which the freeing of hands and the development of the brain created possibilities for complex forms of co-operation and complex forms of collective awareness and control over their behaviour. My analysis of the incest taboo shows that the birth of specifically human social relations would have been impossible without the action of human thought and will mingling their strengths with the strength of objective causes and unconscious factors from which sprang the demand for new forms of social relations. Incest, though at a still more fundamental level, may be compared with the 'invention' of so-called natural languages and the major systems of ideas in the sense that it is the mingled product of unconscious causes and the action of consciousness. Before concluding my remarks on incest, I would, however, like to point out that the universal prohibition of incest entails no particular determination of relations of kinship and provides no explanation of why these should be patrilineal, matrilineal, etc.

I come now to kinship itself. We have seen that the specific feature of relations of kinship is that they engender original social *groups* into which society is divided and which interpose themselves between individuals and their families on the one hand, and society (as a totality requiring to be reproduced) on the other. These social groups form on the basis of a body of abstract relationships that operate as a network; they are centred on the individual yet at the same time off-centre relative to him/her. It is because these relationships are abstract, transitive, and organized as a network, within which some are equivalent and others non-equivalent, that they are capable of reproducing over time, through individuals.

Similarly, because they arise on the basis of a web of abstract, transitive, network-like relationships, relations of kinship have, and still continue to serve as, a framework for the multitude of social functions implying practical and variegated forms of co-operation and sharing. They have provided, and continue to provide, a framework of relations which transcend the necessarily particular character of the practical situations requiring co-operation or sharing. At the same time, though, while bringing people and society together, kinship also divides, and both identical and different interests may clash within it.

One can see why in societies that do not have class relations and state forms, relations of kinship provide an *integrating structure* for society as a whole. Being based on networks of abstract relations running through space and time; binding together by ties of mutual dependence both individuals and the groups to which they belong, yet being broader in scope than the individual family, relations of kinship come gradually to articulate all the different parts of society into a whole capable of reproducing these parts and of reproducing itself. This is why in most classless societies relations of kinship appear to play a *dominant* role. However, in all classless societies one encounters ideas and practices which often operate within relations of kinship, but which all have a wider purpose as well, being concerned with the reproduction of society as such, as a totality, transcending the divisions produced in society by distinctions between kinship groups, the sexes, generations, social divisions of labour, and so on.

This reproduction of society as a whole, as a community, occurs partly through symbolic practices which are politico-religious in essence, entailing expenditures of labour and material resources for the organization of ceremonies, offerings, and sacrifices, and in order to enable the members of society to take part in the ceremonies or other practices. With the emergence of classes or castes, this additional labour which some or all members of society are asked to perform on behalf of the community as a whole, becomes transformed into additional labour to provide the material conditions of existence for a minority that is supposed to represent or serve to a greater degree the community as such, and in order to enable this minority to carry out its politico-religious functions. This additional labour, which everyone was required to furnish from time to time, could have been the starting-point of all the different forms of surplus labour and exploitation of human labour implied in the formation and development of class relations.

As we have seen, once relations of kinship start to operate on the basis of inequality of the sexes and male domination they create conditions for the exploitation of both women and youths. But it is not kinship as such which engenders these relations of domination: it is because these already exist in society that they permeate relations of kinship and press them into the service of their reproduction. If that is so it is pointless to imagine that relations of kinship are capable of becoming anything other than themselves, e.g. class relations. For that to happen the individuals belonging to the dominant sex, for example men, would have to be completely separated from productive work, which would be left in the hands of women and young people, while the men give themselves over entirely to ritual, military or artistic activities. To the best of my knowledge there is no classless society in which male domination is such that none of the men take part in the production of the material conditions of existence.

Ultimately, the question to be answered is how and why do relations of kinship evolve? How? Although they are incapable of becoming anything other than themselves, they do evolve from one form of filiation or another, from one form of marriage and alliance to another. That is how matrilineal systems gradually come to be supplanted by bilineal or cognatic systems, whereas cases of patrilineal systems being transformed into matrilineal ones are practically unheard of. But what drives relations of kinship to change? It looks as if, at bottom, the forces working on them come from somewhere beyond kinship. They are connected with the emergence of new forms of production; with the emergence of new capacities to differentiate and confront the interests of the groups and individuals in society; from the development of new power bases within society. In this respect, relations of kinship are like languages or systems of ideas. They are able to change, but in changing they become changed into other systems of kinship or into other genealogically related languages.

If that is so, it looks as if, of all the functions and components that go to make up a society and that man must reproduce in order to reproduce himself, *the most active* element (in so far as changing itself it exerts the greatest influence in inducing change in the other areas of society) is the body of material, intellectual, and social relations that man produces in order to act upon nature and make it serve his needs. If this hypothesis were confirmed, man's social evolution would appear to be an extension of the evolution of nature, because in nature new species arise when better-adapted mutants are selected in response to changes in the environment.[20] As compared with biological mutations, however, social mutations present a number of specific characteristics; it is with these that I shall conclude this lecture.

The Second Great Moment in the Evolution of Human Society

This analysis thus brings me to the second key moment in the evolution of human society, namely the emergence and development of new social hierarchies, of new relations of domination and exploitation, this time built outside kinship and on the basis of new networks of abstract relationships. This process began over 5,000 years ago, in Egypt or Mesopotamia, and more recently in China, India, or America. It did not occur everywhere and had no reason to do so, but wherever it did happen, we can observe the emergence of social structures having many features in common. We note, for example, the existence of social groups performing distinct functions and entertaining relations of domination and subordination among themselves.[21]

One of the most striking examples of these new societies is the caste system in ancient India which placed the priests—the Brahmins—at the pinnacle of society, followed by the warriors. These two castes lived off the labour of the peasants and

[20] Maurice Godelier, *Marxist Perspectives in Anthropology* (Cambridge, 1977).
[21] Id., *L'Idéèl et le matériel* (Paris, 1984); L. Dumont, *Homo hierarchicus, essai sur le système des castes* (Paris, 1966).

craftsmen, providing them in return with the benefits that the gods can bring and with armed protection. And in this divided society, one structure and one person had the task of integrating all the different parts of society into a whole capable of reproducing itself, and that was the state, incarnated in the king, who alone of all human beings was in a position to perform both the religious and the political functions and to keep each caste and individual in its proper place within the hierarchy of visible and invisible things making up the universe. This hierarchy subsumed relations of kinship, which in this society were subordinated to relations between castes and their reproduction, because castes in India were, and still are, endogamous, each individual being obliged to marry exclusively within his/her own caste and to reproduce his/her caste in reproducing him/herself.

I believe that these new hierarchic relationships first arose when societies found themselves confronted with new situations in which people were forced to negate and renounce some of their traditional ways of thinking and doing and to invent new forms of organization, which must have been regarded as *advantageous* by everyone concerned, even those who came off worst in the process. In my opinion, the stronger of the forces at work in the processes that gave rise to caste and class relations was not violence, but the consent of the dominated to their position. But, for there to be consent, dominators and dominated had to share some common values and representations, and for that the power of the dominators had to be viewed as a *service* which they performed for the benefit of the rest of society, which put everyone else in their *debt*, the only way to honour the debt being to give your wealth, your labour, or even your life, to those performing the services. It was in this process that, for the first time, some men rose above the others and set foot in the immensity that separates us from the gods. Along the way, some of them, such as the Pharaoh or the Inca, ceased to be the representatives of men *vis-à-vis* the gods to become the gods' representatives to men.[22] To penetrate the black box of values and representations held in common by dominators and dominated is surely the most urgent and most difficult task facing the social sciences.

[22] For a critique of Dumont and Meillassoux see Godelier, *L'Idéèl et le matériel*, chap. 7: 'Ordres, castes, classes'.

Thus, like the prohibition of incest and relations of kinship, class, or caste, relations were born simultaneously, both in men's minds and outside. Though dependent on thought for their birth and development, they can never be viewed as just that, nor can they be explained in terms of thought alone. But their birth and development required a formidable evolution in human thought, as Henry Frankfort shows in 'Before Philosophy', when discussing the growth of state societies in Ancient Egypt and Mesopotamia.[23]

We are now better able to see how biological mutations differ from social mutations. Biological mutations occur at random in the individual, and there is no telelogical explanation for them. They survive and serve as a point of departure for the appearance of new species if they happen to be selected by the mechanisms of natural selection whereby life adapts to its environment. Social mutations on the contrary proceed from a mechanism that is more Lamarckian in aspect.[24] They too are born in individuals but they develop within the *relations between them*. They represent a mutation in their relationships, but this mutation is at the same time a mutation in the structures that make up and define a society. These mutations bring about a redistribution of the various social functions that need to be performed in order for society to exist. Man's evolution, therefore, results from a combination of unintentional and intentional forces, which cannot reorganize society without mingling violence with consent. Perhaps today we stand on the brink of a third social mutation, restructuring societies beyond relations of kinship and class relations.

[23] R. McAdams, *The Evolution of Urban Society* (Chicago, 1966); Kent Flannery, 'Interregional Religious Networks', in Flannery (ed.), *The Early Mesoamerican Village* (London, 1976), pp. 329–68; H. Frankfort, *Before Philosophy* (Harmondsworth, 1949); id., *La Royauté et les dieux* (Paris, 1961).

[24] François Jacob, *La Logique du vivant* (Paris, 1970); *Le jeu des possibles: Essai sur la diversité du vivant* (Paris, 1981).

5

Evolutionary Theory: Epistemology and Ethics

BERNARD WILLIAMS

If you consider theories in general, you usually find that if they are capable of very powerful explanatory applications, it is quite hard to apply them vacuously; while if they are easy to apply vacuously, it is hard to make them yield powerful explanations elsewhere. The power to explain, and the possibility of vacuity, are usually related to each other inversely. Particle physics maximizes the first and minimizes the second, while with systems theory, for instance, it is the other way round.

The theory of evolution by natural selection is untypical in this respect. It can be applied in such a way as to yield extremely powerful explanations, but it is also very susceptible to vacuous applications, which explain nothing and barely provide an interesting description of the matters in question. The reason that both these things are possible lies in this, that the basic pattern of explanation that the theory uses is extremely simple and familiar, and it can be applied where it is not obvious—that is why it is powerful—but its strength is very sensitive to certain conditions on the structure of the situation, which control the application of the theory, and it may easily look as though those conditions are satisfied when they are not.

In these remarks I shall consider some applications of evolutionary ideas first in the theory of knowledge and then, more briefly, in ethics. In these areas, as in others that involve human culture, it is very important to distinguish two different questions. The distinction turns on the kinds of thing that are supposedly selected for by the evolutionary process introduced by the theory. This distinction is particularly important with evolutionary models applied to human knowledge. The evolution in question may, first, simply be the evolution that has produced human beings, ordinary biological evolution, and the selection will have been exercised on various species and characters in the emergence of human beings. The question then is: what light can be thrown on human knowledge, belief, understanding, scientific theories, and so forth from our knowledge of the ways in which human beings have evolved? This is the primary emphasis of what has been called 'evolutionary epistemology', as represented notably by the work of Donald T. Campbell.[1] The second interpretation applies a selective model to other items, of a cultural kind, such as scientific theories. These are represented as being themselves selected for. This is the principal emphasis of the work of Sir Karl Popper.

The first of these interpretations sees scientific theories (and so forth) as *affected by* natural selection, namely the natural selection of human beings; the second interpretation sees them as themselves subject to an analogy of natural selection. These two issues, which are very different, are sometimes confused. Thus Franz M. Wuketits has written: 'Since the human mind is a product of evolution . . . the evolutionary approach can be extended to the *products* of mind, that is to say, to epistemic activities such as *science*' (his emphasis).[2] This is simply to run the two matters together. The two approaches are compatible with one another, but they are different. The first approach will claim that certain cognitive capacities in human beings, relevant to science, are indeed the products of natural selection: if so, these capacities will have to be, in some sense, innate. But scientific theories themselves, the concern of the second approach, are not innate to those that hold them—here, the selective pressure is operating on the theories, not on the theorists.

Selection of Theories?

Let us first look at the second approach. Here it is particularly important to make clear the conditions on the substantive application of an evolutionary theory, and to consider what is

---

[1] Donald T. Campbell, 'Evolutionary Epistemology', in P. A. Schilpp (ed.), *The Philosophy of Karl Popper* (La Salle, Ill., 1974), i. 413–51.

[2] Franz M. Wuketits (ed.), *Concepts and Approaches in Evolutionary Epistemology* (Dordrecht, 1984), p. 8.

required by an interesting analogy to natural selection for cultural objects—in this case, scientific theories.

First, we set aside the distinction between phenotype and genotype.[3] In some cases of non-biological evolution, there may be some substantive analogy to this distinction, for instance in the case of firms and their products; but it is irrelevant here. Having taken this step, we then require analogies to three things:

(1) FITNESS (Survival). Fitness is the probability of leaving offspring, so some analogy is needed in the cultural case to leaving offspring. This requires that there should be two 'populations', P_1 and P_2, P_2 later than P_1, and that there should be a given character F present in P_1, which can be present or absent in P_2 partly as a result of its presence in P_1.
(2) SELECTION. Features of the environment of P_1 and P_2 must empirically account for the likelihood of characters present in P_1 being present in P_2.
(3) MUTATION. Something should count as the emergence of a new character.

Condition (1) is in general satisfied for such things as artefacts and other cultural items. Types of conveyance, for instance, once widely spread, at a later time have disappeared. There are further analogies to the biological case more complex than this: thus types otherwise extinct may survive in very special micro-environments. One has to be careful, though, in identifying such examples: thus Bugattis or Delahayes existing in a motor museum do not count, since they do not reproduce there, but are fossils; handmade shirts, on the other hand, are an example.

The problems in applying the analogy of natural selection to such things lie, rather, in the relations between (2) and (3). It is not inevitable that the evolutionary analogy will be fatally weakened at this point, but there is a danger of it. The emergence of some new character or type in the cultural field is usually to be traced to purposive thought by inventors, and so forth; but then the analogy to (3), *mutation*, is very weak, and the structure has low explanatory value. Alternatively an explanation may be found for the emergence of a new character which is nearer to random

---

[3] Equivalently, we treat the survival of the items in question as like the survival of a gene in Dawkins' model.

mutation; thus, for instance, producers often do not know what will succeed, but try out products to see what will succeed. In this case, the analogy to (2), selection, will lie in such things as taste: the product takes on, or it does not. But then a difficulty breaks out about the relations between (2) and (1): they are now, on this interpretation, too close to one another. The facts of survival, and the environmental cause of survival, are more or less the same, that these types of objects are preferred by the public, and the theory once more slides towards vacuity.

How are these analogies to be handled in the particular case of scientific theories?

(1) The relevant characters of theories presumably lie in what they assert, their conceptual features, and so forth. Their presence in a given population, and hence their survival, presumably lies in something like their *acceptance in the scientific community*. There are indeed serious questions about what that notion itself implies. Does it mean, for instance, merely that the theories are used operationally? Or that they are believed to be true?[4] But we will leave that problem to one side.

(2) For Popper, at least, the analogy to selection lies in theories being 'corroborated' or 'falsified'. These are of course, for Popperian theory, *historical* properties, and so they must be, if their impact is to explain the survival that was introduced at (1), since that is itself a social characteristic. But, for the same reason, the fact that a theory has been falsified or corroborated is not in itself enough to explain the disappearance or the survival of that theory in the scientific community. Rather it has to be accepted by the scientific community that the theory has been falsified or, again, corroborated. There are several reasons for this. First of all, and obviously enough, the relevant experiments will have no effect on the scientific community if they are not known. Moreover, whether an observation or experiment is counted as a falsification is itself a function of theory. Moreover, as Kuhn has insisted, it is also a function of what, in terms of the evolutionary analogy, is the mutation rate—that is to say, the emergence of new theories:

---

[4] These questions are often obscured in discussions of the relations of 'our' science to the science of other times. What is it for a science to be 'ours'?

'falsified' theories tend to stay around if there is no better alternative available.

What is the result of combining these various interpretations and concessions? It seems to be, more or less, the claim that theories do not stay accepted in the scientific community if the scientific community has agreed that they are to be rejected. But if this is the outcome, then (1) and (2) are, in this case also, too close to one another, and the account is vacuous.

It is important that on these interpretations the 'selective pressure' and the 'mutation rate' are up to us: they are a matter of what theories are invented, how many experiments are conducted, and so forth. However, nothing in the model could lead to Popper's normative conclusion that we *should* keep up the selective pressure and the mutation rate. (Obviously nothing can be added here by the purely verbal point that it is part of what we call 'science' that these things should be sustained: the question, of course, is about the rationality of what we call 'science'.) If there is a temptation to believe that this normative conclusion does follow from the model of natural selection of scientific theories, or is closely associated with that model, then I think it is due to the error, familiar in the biological field, of supposing that 'fitness' means *more* than the probability of survival or of leaving offspring. The idea will be that the surviving theory will be (also) the *best* theory: in particular, the true theory.

There are indeed many questions about what is meant by the truth of theories. (It is the problem that arose under (1) in connection with the scientific community's 'accepting' theories. It arises, obviously enough, also in relation to (2).) But even if we do move from considering merely the 'survival' of theories as a social fact about the scientific community, to questions of such theories being in some further sense true, it remains quite hard to say anything very illuminating about the effects of selective pressure on the chances of our theories being true. We can certainly say that if we base our beliefs on directed experiment and observation, use the methods of scientific communication, and so forth, we are more likely to end up with true beliefs than if our impressions come from casual observation, hearsay, and similar sources. This is true, and I suspect that, granted the game against nature is not a two-person game, it is necessarily true. Some things will follow from this about how scientific enquiry should be conducted. But not much will follow from it to encourage specific proposals in favour of vigorous scientific competition, repeated attempts to falsify accepted theories, and so forth. Whether these strategies are appropriate, will depend on empirical sociological claims about the effects on the scientific community of different kinds of motivation and styles of scientific culture: to take one example, whether a vigorously competitive atmosphere in science is more likely to lead to honest experiment, rather than to fudging and premature publication. These are well-known and substantive issues, but I do not think that much help with them is to be gained from the natural selection analogy itself. This seems to be a typical case of the phenomenon that I referred to at the beginning of these remarks, that there is only a very narrow area, if any at all, between conclusions that do follow from the analogy, but are merely vacuous, and conclusions that do not follow from the analogy at all.

Science and the Evolutionary History of Human Beings

I now turn back to the first of the two interpretations that were distinguished earlier. Here the evolution in question is the actual biological evolution of human beings. 'Evolutionary epistemology', in a sense appropriate to this interest, emphasizes the continuity of human beings' cognitive processes with those postulated for earlier stages of evolution. Such a theory insists that a theory of knowledge must be a naturalistic endeavour, which needs to be guided by what is known of human evolution. It emphasizes, in particular, that epistemology should be descriptive rather than justificatory, and, in particular, that it should not be foundationalist. A theory of knowledge, moreover, should not imply any 'transfusion of truths from outside'. As Donald Campbell rather colourfully puts it: 'We once saw as through the fumblings of a blind protozoon, and no revelation has been given to us since.'[5]

There is certainly much to be said on broader grounds for a 'naturalised epistemology'; the relation of this to any justificatory

---

[5] Campbell, 'Evolutionary Epistemology', 414.

project in the theory of knowledge is largely beyond the present concern, though I shall touch on some questions of justification later. The present issue concerns the particular contribution made by evolutionary ideas to the broader enterprise of naturalized epistemology.

There is a certain tendency to equate all evolution with learning; in a famous phrase of Konrad Lorenz, 'Life itself is a process of cognition'. This idea is, in itself, the 'evolution of scientific theories' notion taken in the opposite direction: the staphylococcus has phylogenetically learned in developing immunity to a given antibiotic. But this extremely general idea is not the interesting one for evolutionary epistemology, which is concerned with a more specific matter, the capacities possessed by a member of a given species to acquire information about its environment and act directedly in virtue of that information. Neither these capacities, nor the states that they yield, are necessarily conscious. There are very considerable difficulties in distinguishing the required sense of 'acquiring information', from other causal processes that modify behaviour. I shall not try to discuss those difficulties here, but merely note what seems to me a necessary condition on regarding the state of an individual as being, in a sense relevant to these enquiries, a cognitive state: the state invoked in the explanation of behaviour should be the kind of state that can also be, in appropriate circumstances, false.

Campbell and other evolutionary epistemologists tell a hierarchical story, in which the most primitive capacities ('non-mnemonic problem solving') have had successively added to them 'vicarious locomotive devices' (the animal does not have to bang into an obstruction to recognize that it is there), visually and mnemonically supported thought, internalized experiments, cultural accumulation, and science. The effects of this process are, as things typically are in biology, largely cumulative, even though there are many specializations not shared by other genera in the family or order, such as the well-known sonic system of bats.

What can this picture tell us about our exercise of our cognitive capacities, in particular of our special cognitive capacities? One lesson is negative: that we carry with us what has been called in this literature 'a ratiomorphic apparatus', and this apparatus can mislead us. When we recognize this, we are led to a cognitive psychology of error, particularly with regard to probabilities.[6] There are also positive lessons to be learned, for instance in the psychology of perception. Since basic features of our visual system were laid down at an early stage of evolution, we should expect to find certain constant principles of information-processing. This suggests biological contributions to what are indeed biological questions, though they are questions that have complex relations to the judgements made by creatures that are capable of making judgements.

All this is admirable and useful, but it mostly stands at some distance from traditional questions of epistemology. Evolutionary considerations come closest to those questions when they are applied to principles of abstract cognition—when we come to the conceptualization of the world in terms of space, time, and causality, and hence to the general structure of scientific under-standing. Evolutionary epistemologists have believed them-selves to be making contact in this area with traditional problems such as that of innate knowledge, the a priori, and the status of empiricism. Indeed, one of the principal insights of evolutionary epistemology is said to be that 'what is ontogenetically *a priori* is phylogenetically *a posteriori*.' The individual learner does bring something to the world—otherwise he could learn nothing—but it is an inherited accumulation from the evolutionary process.

It is always a problem with these epistemological questions, even in their traditional form, to determine what kinds of properties of the mind are being considered. Wuketits offers as the first postulate of evolutionary epistemology: 'All organisms are equipped with a system of innate dispositions: no individual living system is initially a . . . *tabula rasa*.'[7] The first half of this claim was accepted by classical empiricists who thought that the human mind *was* a *tabula rasa*; they posited at least a disposition to association, and Locke postulated several other mechanisms as well. His denial was that there were innate *propositions* or *ideas*. But that leaves a very severe problem of what counts as a disposition, a principle, a proposition, or an idea. Leibniz, an innatist who was opposed, if anyone was, to empiricism, actually

[6] For important discussions, see D. Kahneman, P. Slovic, and A. Tversky (eds.), *Judgment under Uncertainty: Heuristics and Biases* (Cambridge, 1982).
[7] Wuketits, *Concepts and Approaches*, 5.

agreed with the empiricists and their scholastic predecessors that there was nothing in the intellect that was not in the senses, but added the vital reservation, 'except the intellect itself'.

To distinguish innate from non-innate material in such terms now seems a hopeless task. The question seems to turn on how *specific* or, alternatively, topic-neutral, the innate material is conceived as being. Are human beings general learning machines, or systems with much more specific expectations and models? To give a definite content to the latter account seems to be the most promising and interesting contribution of the evolutionary line of thought to our understanding of human knowledge.[8]

Realism

Evolutionary thought about our cognitive capacities may well have a contribution to make to our understanding of scientific knowledge. But this possibility raises a deep problem. It is not so much a problem *for* evolutionary epistemology as a problem that evolutionary epistemology helps to identify—and it is a virtue of it that it should lead one to identify this problem.

Evolutionary epistemology, obviously enough, assumes the independent existence of a physical world, and tells us quite a lot about its properties, notably just those properties that are invoked in the explanation of how human beings and other species evolved. At the same time, one of the things that it seeks to explain is our conception of that world in those very terms. There is nothing wrong with this situation and, as many writers have recognized, there is no vicious circularity. There would be a circularity if these considerations were used to 'prove' the existence of the external world; that would be to desert the

---

[8] It is certainly more interesting than another line in evolutionary epistemology, which is that of trying to elicit very general principles of learning. These are offered to consciousness from being very general and abstract descriptions of what it is to learn anything. These principles tend to be either vacuous or comically misleading. Thus the fourth hypothesis of Robert Kasper (in Wuketits, *Concepts and Approaches*) states: 'The probability that two or more things will serve the same purpose increases with the number of their common features'. Before one puts in the restrictions that will eventually make this principle into a tautology, it seems to lead to the conclusion that if I can reach something with the help of a tall man, I am more likely to be able to reach it with the help of a short man than with a tall ladder.

naturalistic stance of evolutionary epistemology for foundationalist or justificatory epistemology, and in doing that, it would beg the question.

Evolutionary epistemology is from the beginning a realist theory. The only question is why some of its adherents insist on calling it *hypothetical realism*. Its realism is certainly no more hypothetical than the acceptability of its own assertions. In this sense, as G. Wollmer[9] and others have said, naturalistic epistemology in general represents a true Copernican turn, as contrasted with Kant's.

It is certainly no objection to a theory that its truth is compatible with its own existence. It can even be an attraction of the theory that accepting its truth helps to explain its own existence, or at least the possibility of its existence. But that relationship does put some constraints on what the theory can say. Evolutionary epistemology assigns to various creatures various representations of the world, related to their evolutionary needs and specific natures. Some writers have used in this connection the concept of an *Umwelt*.[10] This concept has rather ambiguous relations to ideas of consciousness: the paramecium, for instance, has an *Umwelt*, but how are things *for* that creature? But, leaving aside the most general questions about the use of this notion, let us allow at least some rather unspecific conception of 'how it is for' a bat or for a cat, and a much less unspecific conception of 'how it is for' us. The theory offers a general scientific vocabulary for explaining these various *Umwelten*, including ours. But that description is our description, and what it expresses is our *Umwelt*. Does this generate scepticism? Might not the cat, from its *Umwelt*, generate in what Thomas Nagel has called its 'furry little mind' a different science?

The first point to make here is that it has not done so. This is a non-trivial fact, and evolutionary epistemology, with its hierarchical story of cognitive capacities, helps to explain why that is to be expected. But then speculation is likely to move to the possible science of possible other, non-terrestrial, advanced creatures. Reflection can, I believe, help to disentangle those

[9] Wuketits, *Concepts and Approaches*, 81.
[10] Introduced, it seems, by Jakob von Uexküll, *Streifzüge durch die Unwelten von Tieren und Menschen* (1934). The term has been happily paraphrased as 'a cognitive niche'.

elements of our representation of the world which we have reason to believe are peculiar to us and perhaps to other terrestrial species, from those that are not. But it is important to emphasize that it will be necessary to do this if the realism implicit in evolutionary epistemology is to be adequately settled, and is not to be exposed to scepticism. Moreover, the resources of evolutionary epistemology itself cannot themselves do this, though they can contribute to it. Evolutionary epistemology must allow some autonomy to abstract scientific theory construction, beyond the local constraints of terrestrial evolution. Otherwise, scientific theory, including the evolutionary theory itself, cannot be anything more than a product of evolution, a contribution to the particularly fantasticated *Umwelt* of one terrestrial species. It is acceptable that it should be a product of evolution, and it is not refuted or shown to be meaningless by the fact that this is what it is; but it must allow for a reflective understanding of the ways in which it is not just peculiarly such a product. In accepting that there is some degree of autonomy to abstract scientific theory in this sense, theorists of evolutionary epistemology will have to recognize that there are limits on what can be achieved within their research programme; moreover, considerable weight is put on the question of how they see the relation of sophisticated scientific theory to evolutionarily more primitive cognitive processes. Those relations are still obscure, and I believe that they are the crux of the question.

Ethics

I turn now, more briefly, to some questions about the relations between evolutionary understanding and ethical outlooks, in particular ethical norms. There are in this area issues analogous to those that arise with knowledge, concerning evolutionarily inherited constraints, and the relations of human culture to the pre-cultural. But there are significant contrasts between the two areas.

The question that has been most often discussed is that of the justification of ethical outlooks from considerations of evolutionary biology, and this question runs immediately into an objection about the relations between *is* and *ought*. I have argued elsewhere[11] that this well-known point does have some force, so far as it goes. If human beings can diverge from some pattern that evolutionary theory supposedly shows to be appropriate to them, then the theory is not going to show that they ought not to. This limited point is sound enough, but the interesting question concerns not so much *ought* and *is*, as *ought* and *can*. The claim that human beings 'cannot' behave or live in a certain way may come in different degrees of strength. At one extreme, it may mean that no single example of any human being doing the thing in question will ever be found: in this sense, a human being cannot lift unassisted a weight of five tons, run a mile on the surface of the earth in one minute, or survive without food beyond a certain period. At the other extreme, it may merely mean that if human beings adopt a norm that requires the behaviour in question, that norm will often be broken, its observance will give rise to a good deal of anxiety, those who comply without anxiety to the norm will be unusual in other respects, and so on. The interesting questions for evolutionary theory and its application to human life lie towards the latter end of this spectrum: the question is whether evolutionary theory can coherently yield *constraints* on social goals, personal ideals, possible institutions, and so forth. If the theory can yield such constraints, then it will necessarily speak at once both to a justificatory and to an explanatory interest. If some biological constraint can rule out, or make unrealistic, some practice or institution, then, if we know about that constraint, this will not only encourage us to decline that practice if it is suggested, but we may also have an explanation of why human communities do not in general display that practice or institution. Similarly, we may wonder whether biological considerations might not also explain the human adoption of other practices, which are conformable to biological constraints.

Any explanation in this style is going to involve the relations between some cultural norm, rule, or institution and a pre-cultural, biological, item. That item is characteristically going to be an inhibition (against mating with siblings, for instance), or a drive (for instance the conditional disposition to defend a

---

[11] In 'Evolution, Ethics, and the Representation Problem', in D. S. Bendall (ed.), *Evolution from Molecules to Men* (Cambridge, 1983). I have used a couple of sentences from this article in the present discussion.

territory). There are certain systematic difficulties that arise in this subject about the relations between such items on the one hand and cultural phenomena on the other; and it is worth recalling, at least in a summary way, what some of these difficulties are. The first and most general question is: if there is a pre-cultural disposition, why should there be a rule? If there is a genetically grounded inhibition, for instance, why is a rule necessary? Is the rule supposed to be an expression of the inhibition? Not every inhibition yields a rule; what are the distinguishing marks of those that supposedly do?

Again there are questions about the content of the drive or the inhibition, as compared with that of the norm or rule. I referred just now to the examples of an inhibition against mating with siblings and a drive to defend a territory; but these are of course external, functionally interpreted, accounts of what these dispositions are. The conceptual content of the descriptions is not supposed to be available to the animals that have these dispositions, nor do they represent the way in which the disposition is operationalized: thus, lacking an ability to recognize its siblings as such, the animal may have an inhibition against mating with other members of the species with which it grew up. The content of the rule, however, represents not the operationalized version of the inhibition, but its functionally described content, and there must be a question about how this rule, with this content, is supposed to have come about.

There may well be answers, in particular cases, to such questions, but the questions need to be pressed. What has to be recognized, and has often been insufficiently recognized by sociobiologists and, still less, by certain popular ethologists, is that there is never any simple mapping from a cultural phenomenon or institution to a biological relationship: parentage, sexuality, and incest have to be understood in cultural terms before they can be related to biological, in particular, genetic considerations. This is not to go beyond ethology: the fundamental ethological fact about this particular species, humanity, is that it lives under culture.

In the case of ethical life, as with human knowledge, the operations of culture must—trivially must—lie within constraints that are set by our evolutionary history and biological inheritance; but culture, in both cases, clearly creates totally new opportunities.

In the case of knowledge, there is the possibility of an autonomy of culture to generate theory that can understand these processes and itself. In the case of ethical life, the autonomy of culture can be seen in the fact that human beings have devised extraordinarily different ways of living, and those ways of living manifestly have to be understood, in good part, in terms of history and the social sciences. The operations of constraints determined by evolution can themselves only be understood if questions are solved associated with the 'representation problem', the problem of the relations between any dispositions described at the biological level, and the cultural phenomena; and to solve that problem requires us to read the historical record.

The opportunities offered by culture can, of course, be negative. Lorenz said that 'It was a privilege of man to believe pure nonsense',[12] and that privilege has certainly been exercised in the failure even to try to read the historical record in many of these connections—for instance, in seeking to understand relations between men and women. Sometimes the nonsense has consisted in trying to evade biological constraints, to live according to illusions that human beings are not animal or corporeal. To get rid of those illusions, it is certainly necessary to regain contact with biological aspects of our life, and, in the most literal sense, simply to know better what we are. But it is equally an illusion to suppose that there is any route to understanding better what we are that could be found simply on the basis of biological information, and did not require insight into culture.

[12] Quoted from a lecture course, by R. Riedl in *Biologie der Erkenntnis* (Hamburg and Berlin, 1981); trans. as *Biology of Knowledge* (Chichester and New York, 1984).

6

Evolution in the Arts: The Altar Painting, its Ancestry and Progeny

E. H. GOMBRICH

'Evolution in the Arts' is the title of a very learned book of 1963 in which Thomas Munro, a distinguished authority on aesthetics, surveyed the vast literature on the subject, including, of course, the theories of Herbert Spencer in more than 500 double-columned pages.[1] In reviewing the book[2] I ventured to criticize the author for his benevolent neutrality towards all varieties of this multi-faceted theory which left one wondering what he really thought. Perhaps it was my uneasy conscience which prompted me to accept the invitation to deal with this elusive subject in this series, knowing full well that I have forfeited my right to be heard with benevolent neutrality.

For to confess it without further ado, I have myself been toying with certain analogies I thought to perceive between evolution in nature and evolution in the arts. Indeed, when I was asked at pistol-point by my French publisher what to call a collection of my essays which he had already printed, I plumped for the title 'L'Ecologie des images'.[3] The title caused a certain flutter in the critical dovecots, particularly in France and in Italy, though I had meant nothing more sensational by it than the idea (which I had quite frequently put forward before) that what we call changes in style can be interpreted as adaptations, on the part of the working artists, to the functions assigned to the visual image by a given society. I say advisedly of the artists, not of art, because I am an individualist in such matters and prefer to see the history of art as the resultant of living people responding to certain expectations

---

[1] Thomas Munro, *Evolution in the Arts and Other Theories of Culture History* (The Cleveland Museum of Arts, Cleveland, 1963).
[2] *The British Journal of Aesthetics*, 4 (1964), 263–70.
[3] Paris, 1983.

and demands which, in their turn, they may also help to stimulate, or at least to keep alive. In fact the metaphor of the ecological niche appealed to me precisely because it does not imply a rigid social determinism. The study of ecology, I believe, has alerted us to the many forms of interaction between the organism and its environment which renders the outcome quite unpredictable.[4]

Needless to say, these generalizations do not only apply to the arts of the image makers. All the specialized skills that together make up the fabric of civilization must have evolved in answer to demands which fed on their satisfaction. But in the history of technology and of science these demands are perhaps more easily specified than they are in the history of the arts. Whether we think of aviation or medicine it is not difficult to explain the driving force behind their evolution.

To apply the term evolution to all the arts indiscriminately would seem to imply that like the arts of flying or of healing the arts of sculpture and of painting were always intended to serve a given statable purpose. Up to a point, I think, this was Spencer's view. He saw the arts as a means of producing 'aesthetic enjoyment'. I would not deny that there is such a thing, but it has always seemed to me one of the pitfalls of philosophical aesthetics that it so easily leads to tautologies. You might say that neither cave paintings nor minimal art would have been produced if these products had never pleased anyone. I forget who it was who said in this context: 'For those who like that kind of thing, it is the kind of thing they like.'

My book on *Art and Illusion*[5] contains a chapter entitled 'Reflections on the Greek Revolution' in which I tried to be a little more precise. I there suggested that the slow but steady approximation of Greek art to natural appearances, the evolution of mimesis, might have been due to the same demands that also resulted in the development of the drama, the demand, or perhaps the craving, to see the events told by the poets and historians re-enacted as if they were happening in front of us—a demand I later condensed into the formula of the 'eye-witness principle'.[6] Applying this principle must ultimately have led to the mastery of

[4] See my essay on Hegel in *Tributes* (Oxford, 1984), p. 64.
[5] New York and London, 1960.
[6] See my volume *The Image and the Eye* (Oxford, 1982), s.v. in Index.

foreshortening, the rendering of a unified space, and of light and shade. To illustrate these conquests it suffices to contrast the famous mosaic of Alexander's victory over Darius (Pl. II) with an ancient Egyptian victory monument in which Thutmosis III is seen holding and smiting a bundle of Asiatic prisoners (Pl. I). By way of radical simplification I'd call this style pictographic and that of the Hellenistic battle-piece photographic.

Admittedly this account of the function of the image in classical antiquity is very one-sided. The techniques of image-making never serve the same kind of statable purpose as does the technology of aviation or medicine. I still believe that dramatic narrative was a dominant aim of ancient art, but obviously the older and more persistent demand must have been for the creation of cult images. After all, the fame of Phidias rested on his colossal statues of the Olympian Zeus and the Athena Parthenos. Both these works satisfied another demand, they must have been immediately recognizable as the divinities they represented, and this recognition also depended on their exhibiting the correct traditional attributes. One copy of the cult image of Athena—artistically not the most attractive one—is at any rate complete, showing her traditional attributes such as the Gorgoneion, the formidable trophy she wears around her neck, the shield and spear, and the statue of Nike her auspicious helpmate (Pl. III). Many ancient statues we see in our museums were found without their attributes which were sometimes dubiously restored, but we can envisage them from other images, such as that of Poseidon with his trident and a fish on a fifth century amphora (Pl. IV).

In a sense these attributes are the distinguishing features which serve to identify the God; ultimately they, too, are often derived from a narrative context, from the role attributed to the God in mythology, where Poseidon shakes the earth with his trident or Apollo plays the lyre. What history tells us, moreover, is that the increasing dominance of this narrative function also began to affect the isolated image. Praxiteles represented Aphrodite for her shrine at Knidos as emerging from her bath, and Lysippus may have invented the image of Herakles resting on his club from his labours, preserved in the Hercules Farnese.

I believe that when it comes to the discussion of Christian art in the middle ages it is even more important to keep in mind the multiplicity of function the image was expected to serve or, indeed, to avoid, and it is to the problem arising from these divided aims that I should like to devote this lecture. Speaking again schematically it may be said that the Church had insisted on replacing the dramatic or evocative function by the pictographic one. According to Pope Gregory the Great painting was to serve the illiterate laity for the same purpose for which the clerics used reading.[7]

Ever since Ruskin this didactic purpose of the image in Christian art has been emphasized, and though one may doubt whether those who could not read texts could in fact read the signs and symbols of ecclesiastical art, there is no denying the element of pictographic clarity of this tradition that culminated in the imagery of the great cathedrals. I need only remind you of their porches where the sacred personages stand in serried ranks, as in Chartres, on the northern transept begun in 1194, allowing us to read off their identities from these abbreviated symbols of their role (Pl. V)—Melchisedek with his chalice and the censer of a priest, for he offered Abraham bread and wine, Abraham whose sacrifice of his son Isaac is here compressed into a pillar while he is looking up to the angel of deliverance, Moses with the tables of the law and the pillar of the brazen serpent, Samuel with a sacrificial lamb, and King David with crown and spear.

Or remember the great stained glass windows of the same cathedral, one of which shows the ancestry of Christ, the so-called Tree of Jesse in almost diagrammatic form (Pl. IX).

In calling the style arising from this function pictographic, I am not, of course, making an aesthetic judgement, remembering how many great works of art testify to its artistic potential. Let me contrast two representations of the Last Supper, both taken from the screens of medieval cathedrals. The one from Modena (Pl. VI), dating from about 1184,[8] the other from Naumburg (Pl. VII),[9] created about eighty years later by one of the greatest German

---

[7] 'ut hi qui litteras nesciunt, saltem in parietibus videndo legant quae legere in codicibus non valent', *Sancti Gregorii Magni Epistolarum Lb. IX*, Epist. CV, Migne, *Patrologia Latina*, XVII, cols. 1027–8; 'Nam quod legentibus scriptura, hoc idiotis praestat pictura cernentibus', ibid. *Epist. Lb. XI*, Epist. XIII, col. 1128.

[8] For dating and bibliography see the exhibition catalogue *Lanfranco e Wiligelmo, Il Duomo di Modena*, ed. Enrico Castelnuovo *et al.* (Modena, 1984), pp. 560–3.

[9] Ernst Schubert, *Der Naumburger Dom* (Berlin, 1968).

masters. The earlier work is a lapidary illustration of the words of the gospel of St John, where Christ says:

Verily, verily, I say unto you that one of you shall betray me. Then the disciples looked one on another, doubting of whom He spake. Now there was leaning on Jesus' bosom one of his disciples, whom Jesus loved, Simon Peter therefore beckoned to him that he should ask, who it should be of whom he spake . . . Jesus answered, He it is, to whom I shall give a sop, when I have dipped it. And when he had dipped the sop He gave it to Judas Iscariot . . .

The text is illustrated with the utmost economy. The central group shows us the disciple leaning on Jesus' bosom, and the gesture of Jesus as he gives the sop to Judas. We must look a little more closely for St Peter who had beckoned that Christ should be asked, but there is only one of the disciples who has raised the hand in a speaking gesture, while all the others merely look at their neighbours. Remembering, as we all do, Leonardo's evocation of that scene we may find it hieratic, but it is all there.

In Naumburg we are well on the way to Leonardo's dramatic evocation. Note the famous trait of Jesus holding his sleeve while handing Judas the sop, so that it should not get into the dish in front of him, not to speak of the freedom with which each individual figure of the drama is realized. Here as elsewhere I hope to be forgiven if I use these works of art merely briefly as an illustration of a theoretical point; the point that the difference between these two works cannot be expressed merely in terms of the evolution of formal means. I have often said that I believe Emile Mâle[10] was fundamentally right when he reminded us of the new approach to the sacred story that can also be documented from popular sermons and miracle plays all of which strove to evoke the events as vividly as possible, thus creating new expectations and a novel demand somewhat analogous to what I have called the Greek revolution.

Remembering that the Naumburg group dates from about 1260 it is worth recalling that this is also the period to which the first historian of art, Giorgio Vasari, assigned the beginning of the Renaissance. Knowing nothing of the North he saw it embodied in

[10] *L'Art religieux de la fin du Moyen-Âge en France* (Paris, 1908), see also my 1976 Walter Neurath Memorial Lecture, *Means and Ends, Reflections on the History of Fresco Painting* (London, 1976).

the sculpture of Nicolò Pisano who had profited in his evocations from his study of ancient art.

If I am asked why painting lagged behind sculpture in meeting the demands which I have postulated, I would reply that certain media are less adaptable to certain requirements than others. It is far from obvious for the painter how to create the illusion of really looking into the room where the Last Supper is taking place. Even Giotto did not quite succeed. He lacked the scientific tool that has to be evolved by trial, error, and scientific calculation in the slow progress that you find described in any history of Italian art from Giotto to Tintoretto.

Far from wanting to go over this ground again I should like to concentrate instead on the inadequacy of the 'eye-witness principle' for the explanation of many paintings of the period, including some of the most famous ones. For, like their ancient predecessors, Italian artists also had to meet the demands of a very different task—the creation of cult images. Now it is true that the doctrine of the Latin Church excluded the worship of images, but it could not exclude the popularity of cults, the cult of saints and their relics and the development of a genre of art to which I should like to devote most of the remainder of this lecture—the decoration of the altar.[11]

The emergence and evolution of this art-form demonstrates to my mind the way a great variety of independent factors interact, ultimately to produce results which could never have been predicted at the outset. It is more than a pun if I refer back to the term of the ecological niche, because it allows me to explain why this niche had remained empty during the first millennium of the Christian era.

One of the facts here was purely liturgical: we have been reminded in recent years after the Vatican Council that in the early centuries the priest stood behind the altar, facing the congregation, as we see on a famous Carolingian ivory in Frankfurt (Pl. X). While this practice, which has been revived in recent years, persisted, the mensa, the altar, was really a table, and nothing could have been placed upon it without obscuring the officiating

[11] The masterly essay by Jacob Burckhardt, 'Das Altarbild', in *Beiträge zur Kunstgeschichte von Italien*, published posthumously in 1898 (*Gesamtausgabe* 12 (1930)) has never been superseded, but it only deals with one period and school.

priest. All that could be done to decorate the altar was to adorn the front of the table, and indeed there were such altar frontals, the most famous being the golden antependium, now in the Musée Cluny, from the early eleventh century, given by the German Emperor Henry II to the cathedral of Basle (Pl. VIII). We see the diminutive figures of the imperial donors crouching at the foot of the majestic central figure of Christ who is accompanied by the three archangels, and St Benedict with the crozier and the book. There are also less costly frontals or antependia with pictorial decorations, but the demand for such images only became insistent when the priest changed his position before the altar while celebrating mass, as shown in a fifteenth-century Flemish miniature (Pl. XI), where the priest is seen elevating the host in front of a large altar structure. The type is still extant in the North—for instance the magnificent shrine by Hans Brüggemans now in Schleswig, dating from 1521 (Pl. XII). I trust no one will need persuading that such an elaborate work stands at the end of a long and complex evolution which I can only present in a radically simplified form, a kind of diagram which can only relate, at best, to the real course of events as the London Underground map relates to the actual network of lines.[12]

The wooden triptych represented as standing on the altar in the background of Sassetta's painting with the Funeral of St Francis in the London National Gallery illustrates the earlier form these church furnishings frequently took in the Italian South (Pl. XIII). They offered a visual display for the congregation during the service and made the altar into the focus of devotion for anyone entering the church. But while the change in the position of the priest made this development possible, it can hardly be described as a sole cause. Another element enters this momentous development which brings it home to us that evolution in the arts can never be studied in isolation from geographical and historical factors. I speak of the role of the image in the neighbouring lands dominated by the Eastern, the Byzantine Church.

I hope I may steer clear of the complexities and horrors of that issue, horrors because of the bloodshed it caused when the

---

[12] For this and the following, see Helmut Hager, *Die Anfänge des italienischen Altarbildes; Untersuchungen zur Entstehungsgeschichte des toskanischen Hochaltar retabels*, Veröffentlichungen der Biblioteca Hertziana (Munich, 1962).

iconoclastic movement nearly tore the Byzantine Empire to shreds. Suffice it to say that after that terrible crisis the image or icon was accorded a special status in the Greek Church which largely removed it from the innovations and experimentations which characterize developments in the Latin West.

But what is important in our context is that the Icon is not an altar painting. Its place is on the *iconostas*, the screen that separates the laity from the choir. Now the history of the *iconostas* may in some respect run parallel to that of the Latin altar,[13] but this need not concern us. What concerns us is a historical event which marks an important point in my diagram. I mean the conquest of Constantinople by Latin crusaders in 1204.

The exact significance of this event may be a matter for learned debate,[14] but it can hardly be an accident that this temporary removal of the political barrier between East and West coincided with a vogue of panel paintings in Italy which can only be described as imitation icons. The great seaport of Pisa seems to have been one of the main centres of import and soon also of the production of such easily portable images as the painting of the Madonna now in Santa Maria del Carmine in Siena (Pl. XIV) which was probably painted in Pisa and so clearly reflects a Byzantine type. Indeed, seen from Byzantium these incunables of Italian panel painting must have looked like provincial or even folk art.

It was this type of painting derived from Byzantium which Vasari had in mind when, in the sixteenth century, he looked back on the evolution of painting in Italy and described what he called the crude and ugly manner of the Greeks. He thought that this style had dominated all the centuries since the decline of antiquity and that it was only due to his compatriot, the almost legendary painter Cimabue, mentioned by Dante, that Italian painting began

[13] See Hager, op. cit., 66–74.

[14] Among recent discussions of this complex connection I should like to quote Kurt Weitzmann, 'Crusader Icons and Maniera Greca', in Irmgard Hutter (ed.), *Byzanz und der Westen, Studien zur Kunst des europäischen Mittelalters* (Sitzungsberichte der österr. Akademie der Wissenschaften, Phil.-Hist. Klasse, 432; Vienna, 1984), pp. 143–70; Ernst Kitzinger, 'The Byzantine Contribution to Western Art of the Twelfth and Thirteenth Centuries', in Ernst Kitzinger, *The Art of Byzantium and the Medieval West* Bloomington, Ind., 1976), pp. 357–78, and Otto Demus, *Byzantine Art and the West* (New York, 1970), all with further bibliography.

I. *Tutmosis III Smiting the Asians* 15th century BC

II. *Alexander's Victory over Darius*, mosaic after a Hellenistic painting, *c.*100 BC

III (*above, left*). *Athena Parthenos*, Roman copy after a statue by Pheidias, *c*.447–432 BC

IV (*above, right*). *Poseidon*, 5th century: From the Panathenaic Amphora by the Kleophrades painter

V (*right*). *Patriarchs and Prophets*, *c*.1200: From the porch of the north transept, Chartres Cathedral

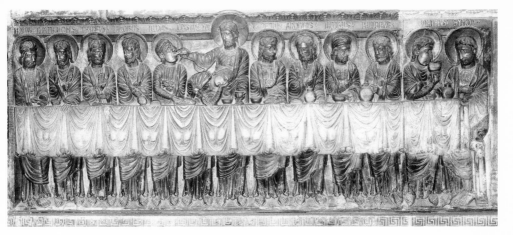

VI. *Last Supper*, *c*.1180: From the choirscreen of Modena Cathedral

VII. *Last Supper*, *c*.1260: From the choirscreen of
Naumburg Cathedral

VIII. *Antependium of Henry II*, early 11th century

IX (*left*). *Tree of Jesse*, stained glass, 12th century: From the west façade, Chartres Cathedral

X (*above*). *Priest Celebrating Mass*, ivory, 9th–10th century

XI. *Priest Celebrating Mass*, miniature, Flemish 15th century

XII. Hans Brüggemans, carved polyptych, made in Bordesholm, 1521

XIII. Sassetta (Stefano di Giovanni), *Funeral of St Francis*, panel from an altar-piece, 1444

XIV. *Madonna*, probably painted in Pisa, 13th century

XV. *Icon of St George*, before 1200

XVI. Bonaventura Berlinghieri, *St Francis*, 1235

XVII. Bonaventura Berlinghieri, *Sermon to the Birds* (detail of XVI)

XVIII. Giotto (attributed to), *Sermon to the Birds*, *c*.1300

XIX. Giotto (attributed to), *St Stephen*, *c*.1300

XX. Vitale da Bologna (active 1334–1359),
Adoration of the Magi and Saints

XXI. Taddeo Gaddi, *Altar Painting*, mid-14th century

XXII. Nardo di Cione (Orcagna), *Altar Painting*, 1357

XXIII. Nardo di Cione (Orcagna), *Christ in the Storm*, Predella (detail of XXII)

XXIV. Domenico Veneziano, *Sacra Conversazione*, c.1440

XXV. Domenico Veneziano, *Predella* (detail)

XXVII. Lorenzo Costa, *Bentivoglio Madonna*, fresco, 1488

XXVI. Jacopo del Casentino, *Madonna and Donors*,
mid-14th century

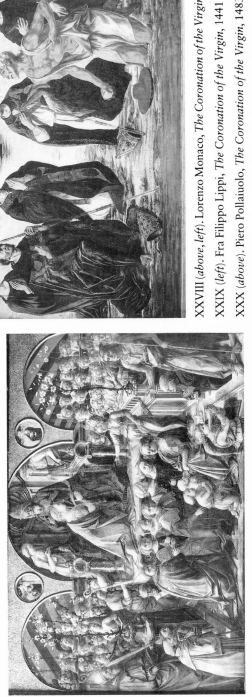

XXVIII (*above, left*). Lorenzo Monaco, *The Coronation of the Virgin*, 1414

XXIX (*left*). Fra Filippo Lippi, *The Coronation of the Virgin*, 1441

XXX (*above*). Piero Pollaiuolo, *The Coronation of the Virgin*, 1483

XXXI. Rogier van der Weyden, *Deposition*, altar painting, *c.*1435

XXXII. Rogier van der Weyden, altar
wing, middle of the 15th century
(detail)

XXXIII. Master of the St Bartholomew
Altar, altar wing, early 16th century

XXXIV. Geertgen tot Sin Jans, *Tree of Jesse*, c.1480

XXXV (*left*). Leonardo da Vinci, *St Anne*, *c*.1508
XXXVI (*right*). Leonardo da Vinci, *St Anne*, *c*.1502, cartoon

XXXVII (*left*). Luca di Tommè, *Virgin and Child with St Anne*, second half of the
14th century
XXXVIII (*right*). Benozzo Gozzoli (d. 1497), *Madonna and Child with St Anne and Donors*

XXXIX (*left*). Titian, *Polyptych with Resurrection*, 1522

XL (*below, left*). Titian, *St Mark Enthroned, c.*1511

XLI (*below, right*). Titian, *St Sebastian* (detail of XXXIX)

XLII (*left*). Titian, *St Sebastian*, drawings, 1522
XLIII (*centre*). Titian, *St Sebastian*, drawing, 1522
XLIV (*right*). Michelangelo, *Slave*, destined for the tomb of Julius II, *c*.1516

XLV. Teniers the Younger, *The Collection of Leopold Wilhelm*, *c*.1650

to bestir itself, after which Giotto set it on its path of unbroken progress leading to the perfection of Raphael.

It is a neat story, but the events can also be read somewhat differently. It may be no less correct to say that Raphael's Madonnas, like the statue of Aphrodite by Praxiteles, testify to the triumph of the narrative mode over the tradition of the symbolic cult image, with further consequences to which I shall turn at the end of this lecture.

Not that the duality of pictographic symbolism and narrative representation can only be found in Western art. A twelfth-century icon from Mount Athos shows St George with his attributes surrounded by narrative scenes from his life (Pl. XV).[15] One of the earliest Italian panel paintings devoted to St Francis of Assisi, the panel by Berlinghieri in Pescia near Pisa (Pl. XVI), likewise shows the saint with his emblematic stigmata surrounded by episodes from his legend. We art historians like to contrast this scene of the Sermon to the Birds (Pl. XVII) with the same scene from the fresco cycle of Assisi (Pl. XVIII). Whether or not it is by Giotto, it again illustrates to perfection the gulf between a realistic evocation and an almost pictographic rendering.

But such a comparison almost begs the question, the question of the purpose and function of these images. We learn more about the clash of tendencies by looking at a wing of an altar, attributed to Giotto. It represents St Stephen in the garb of a deacon and at the back of his tonsured head a small stone which serves as the attribute, a reminder of his martyrdom by being stoned to death (Pl. XIX). A typical altar by Giotto's follower Taddeo Gaddi (Pl. XXI), of around 1340, illustrates the context of this conventional form, a series of panels, no longer quite in their original setting, with the Holy Virgin in the centre, flanked by the Baptist, with his pointing gesture, 'ecce agnus Dei', St Lawrence with the gridiron, and on the opposite side again St Stephen, and St James. We are so used to these divergent modes of painting that we sometimes forget that to ignore these inherent tensions can cause us to misunderstand the meaning of certain images.

There is a charming panel in the Edinburgh Gallery attributed to the fourteenth-century master Vitale da Bologna (Pl. XX). It is described in the catalogue as representing the Adoration of the

[15] Demus, *Byzantine Art*, p. 212.

Magi. To be sure, we see the three kings of legend, representing, as they traditionally do, three ages of man. The old king has deposited his gift of gold in the form of a chalice onto the seat of the Virgin, the second, in his prime, who points to the star, holds the censer, and the youth presumably carries myrrh in his box or coffer. Below, in the right-hand corner, we see the horses from which the travellers have dismounted, with their grooms. They wear exotic headgears to show that they have come from distant lands though one of them is only visible behind the horse. But if this part of the panel justifies the interpretation of the catalogue as a narrative the two female figures on the other side do not. They are two female saints, and again, if we look closely, we discover other figures presumably of further saints recognizable only by their haloes and the crown on one of their heads. Whoever they are, they are not part of the narrative, the artist did not imagine or envisage them standing by the side of the scene when the Magi arrived from the East. The one holding a wheel is of course St Catherine, the one with the sword in all probability St Ursula. They are again marked by their attributes or emblems which remind the faithful of their characteristic martyrdom. Should we then really describe the panel in Edinburgh as *The Adoration of the Magi*? Is it not rather to be interpreted as an assembly of saints, namely Caspar, Melchior, and Balthasar together with female saints? The question may be an idle one, except that it brings it home to us that what may have mattered to the faithful was the presence of these intercessors rather than their role in a particular sacred episode.

I believe that in nature hybrids often prove infertile, and in the history of art, too, this rather confusing mixture of modes was not granted a long life. What we observe instead in the history of altar painting is a tendency to separate the two modes, assigning one part of the altar to the symbolic composition and another subsidiary one to the narrative. Thus the great altar-piece by Orcagna dedicated by the Strozzi to Santa Maria Novella exemplifies this solution (Pl. XXII). On the main panel we see Christ enthroned within a mandorla of angels handing the keys to St Peter and a book to St Thomas Aquinas. Needless to say this is not the evocation of a scene which was imagined to have taken place in a here and now, but a purely symbolic visualization of the role assigned to these saints, just as the other sacred personages, all of them not only marked with their emblems but labelled with their names, are here assembled as if they were witnesses to the solemn act, the Virgin protecting St Thomas, St Catherine with her wheel and St Michael with the dragon on one side, and opposite the Baptist behind St Peter, St Lawrence with the gridiron, and St Paul with his epistle to the Romans, neatly sealed.

The late Millard Meiss in his book *Painting in Florence and Siena after the Black Death*[16] sensitively analysed this great composition which for him represented a departure from the realistic tendencies of the earlier generation of Giotto and a return to a more medieval and hieratic style. I would not contradict him, but we must not overlook the functional aspect of the altar painting, for underneath, in the predella, the artist has given narrative its full scope with legends from the saints represented above (Pl. XXIII). Thus the scene of Christ and St Peter on Lake Genezareth is rendered with great dramatic intensity—remember the moment of Matthew 14, when St Peter, having left the boat and walked on the water finds himself sinking and cries out, 'Lord save me', and Christ replies, 'O thou of little faith'.

In the next century, in the art of the quattrocento, the story-telling conception gains even more ascendency over the symbolic or pictographic one. It was this shift in emphasis which led to a new form of altar painting as exemplified in Domenico Veneziano's wonderful St Lucy altar-piece (Pl. XXIV), so typical of the second quarter of the fifteenth century. The arcading of the hall still reflects the earlier tradition, but St Francis, St John the Baptist, St Zenobius the bishop, and St Lucy are shown together in the airy chapel into which the light is streaming, where they seem to be paying homage to the Virgin on her throne. The togetherness of the sacred personages has led to the type of composition being referred to as a *Sacra Conversazione*, a term of the nineteenth century, but of course there is no conversation in any earthly sense going on, not an event but a symbolic assembly. Here too the narratives of events were confined to the predella, I say *were*, because such small panels were easily taken away and two are now in Washington, and two at the Fitzwilliam Museum in Cambridge. Each shows a miracle from the life of the saint represented above,

[16] Princeton, NJ, 1951.

such as the miracle of St Zenobius restoring a child to life after an accident (Pl. XXV), and I need not enlarge again on the dramatic character of this narrative which seems to be laid in a typical Florentine street rendered with convincing perspectival skill, perspective being the form most suited to what I have called the evocative function.

The feature of altar painting which was most strikingly affected by the novel demands of a unified spatial setting was the place of the donor in the composition. 'Compare and contrast' as the examiners say, an altar painting by Jacopo del Casentino (Pl. XXVI) from the early fourteenth century with Lorenzo Costa's altar of 1488 (Pl. XXVII). Both are dedicated to the Holy Virgin enthroned in the centre, but in the early work the donors who paid for the altar and hoped for reward in heaven are shown diminutively small in comparison with the holy personages, as on the altar frontal from Basle. In Costa's painting the whole family or clan of the Bentivoglios are crowding in, no different in scale from the Madonna. It is tempting in all such cases to regard this change as a symptom of a changed mentality, the 'self-assertiveness of Renaissance man' as compared to the humility of the medieval donor. But I am not sure that such an inference would be justified. Costa worked within a different pictorial tradition which excluded a disregard of scale and we can only admire the skill with which he managed nevertheless to respect proprieties and to preserve some kind of psychological distance between the heavenly and the mundane sphere by purely compositional means.

For there is no denying that the expectations aroused by the novel achievements of realistic painting confronted the producers of altar paintings with special problems. Just as Costa felt no longer free to manipulate the size of the figures to express their humility or grandeur, so the needs of the genre could get into conflict with other progressive achievements. The ends always influence the appropriate means and the function of the altar painting demands that it should be clearly visible from afar.

In a recent volume of my essays[17] I juxtaposed the illustrations of two Florentine quattrocento altars, both representing the Coronation of the Virgin, one by Lorenzo Monaco dating from 1414 (Pl. XXVIII), the other by Fra Filippo Lippi dating from

[17] *New Light on Old Masters* (Oxford, 1986), p. 102.

1441 (Pl. XXIX). My purpose in that book was to show how Fra Filippo's mastery of realistic skills had to some extent jeopardized the spiritual quality of the earlier Gothic evocation of the heavenly ceremony; in addition it also resulted in a painting that must be inspected at close quarters to reveal all its subtlety and beauty, while the Lorenzo Monaco proclaims its meaning at one glance from a distance. Two Florentine painters of the next generation, Piero Pollaiuolo (Pl. XXX) and Sandro Botticelli, showed in their altar paintings of the same subject that they had absorbed the lesson and adjusted their means to the specific task. Though they could no longer work with a golden background, they still achieved that lucidity of composition which the genre of the altar painting demands.

That the problem is not one we construct from hindsight, but is inherent in the task, can be shown by its very fertility. Both the artists of Italy and those of northern Europe sensed its urgency and worked out ways to resolve these tensions.

Rogier van der Weyden's great Deposition (Pl. XXXI) roughly dates from the same period as Filippo Lippi's Coronation of the Virgin, but it combines the most moving psychological realism with the refusal to create a pictorial illusion of reality, his group is assembled in a simulated sculptural shrine. Or take his panels of about 1445, now in Philadelphia, in which the golden background is confined to an upper zone. Before that we find a stone wall draped with what may be called a cloth of honour to set off the figures of the Saviour, of the Virgin and St John (Pl. XXXII), again painted with all the realism without destroying the sense of distance. It needs the tact and inspiration of a great master to achieve this balance between conflicting demands.

Towards the end of the century the German Master of the St Bartholomew Altar-piece also used the device of a cloth of honour to set off the donor and the saints (Pl. XXXIII), but he could not refrain from showing the peak of mountains behind the cloth, a touching display of illusionistic skill, but slightly incongruous.

I hope to have shown that it might be fruitful to look at the evolution of altar painting in the light of the problem inherent in the task, the need of harmonizing the requirements of traditional didactic symbolism with the means of realism. I know no more telling example of these inherent tensions than the painting by

Geertgen tot Sint Jans from the late fifteenth century (Pl. XXXIV). We see a large group of realistic modishly dressed figures balancing on the branches of a tree like acrobats. It is of course a representation of the Tree of Jesse, the family tree of Christ, with David playing the harp in the centre and other ancestors precariously perched aloft. When we remember the twelfth-century stained glass window from Chartres cathedral illustrating the same theme (Pl. IX), we appreciate the ease with which a non-realistic mode of representation could present this visual symbol.

This is an extreme example, and I would not have shown it at this point if I did not think that its lesson has a bearing on one of the great altar paintings which is now housed in the Louvre, Leonardo's *St Anne* (Pl. XXXV). I have suggested in my book *Symbolic Images*[18] that the problem confronting Leonardo can only be explained in terms of these two conflicting traditions or functions. A polyptych of the trecento by Luca di Tommè illustrates what I mean (Pl. XXXVII). Here there is no doubt in anybody's mind what we are supposed to see. Each of the saints is marked by their symbolic attribute, and St Anne is also so marked by having on her lap the Holy Virgin; how do we know she is the Holy Virgin? Because she holds on her lap the Christ Child; the picture is in fact a diagrammatic family tree. But with the coming of realism the conflict of which I have spoken made itself felt. In Gozzoli's version of the theme, the inconsistency of scale reflects the inconsistency of aims (Pl. XXXVIII). Leonardo could not possibly be satisfied with such a half-hearted solution, and we know from his sketches and his versions how hard he worked to reconcile the irreconcilable and to weld the whole into a group which is both an evocation and a symbolic presentation.

The cartoon in the National Gallery in London is unintelligible except as an attempt to solve this problem (Pl. XXXVI). Why is the Virgin seen half sitting on the lap of her mother? Clearly because if she were to slide down and sit beside her, the older woman would be taken for the mother of little St John, that is, for St Elizabeth, not St Anne. Leonardo later eliminated St John, the cause of this extra complication. He, if anyone, conceived of the art of painting as an evocation and re-creation of reality, and so great was his fame as a painter that when Filippino Lippi, who had

[18] London, 1972, p. 16.

been commissioned to execute an altar painting of St Anne, heard that Leonardo, just back from Milan, was ready to undertake it, he stood down to make way for the renowned artist. Vasari, who tells us this story also relates that when the cartoon for that composition was ready in Leonardo's studio 'not only were artists astonished, but the room where it stood was crowded with men and women for two days, all hastening to behold the wonders produced by Leonardo; and with reason,'—Vasari adds—

for in the face of the Virgin is all the simplicity and loveliness which can be conceived as giving grace to the Mother of Christ as she contemplates the beauty of her son whom she holds in her lap. St Anne looks upon the group with a smile of happiness, rejoicing to see her earthly offspring becoming divine.

It seems that Leonardo himself was not satisfied with this generalized evocation, and that he gave visitors a more detailed theological explanation of the symbolic interaction of the figures with the Virgin wishing to remove the child from the lamb, the sacrificial animal and her mother restraining her so as not to hinder the sacrifice for the salvation of mankind.

What matters to me in this historical account is another step in the evolution from the cult image to a work admired for its own sake as a work of art. There is no evidence that Leonardo's painting which he took with him to France and which is now in the Louvre ever stood on an altar.

I do not want to overstate my case: I do not claim that earlier works of art, whether Western or Eastern, had not also aroused admiration and that their masters were not also intent on displaying their skill for its own sake or to give aesthetic pleasure. But I believe it is significant that in the early sixteenth century, the period we call the High Renaissance, the general interest in the work of the artist had grown to such a pitch that even altar paintings were predominantly seen as opportunities for the display of mastery. When, around 1518, Giuliano de Medici commissioned two altar paintings for his cathedral at Narbonne from two famous painters in Rome, Raphael and Sebastiano del Piombo, the occasion was watched as a competition between two rival cliques. Sebastiano was the protégé of none other than Michelangelo, and no love was lost between his friends and those of Raphael, the darling of Pope Leo X. In order to secure Sebastiano's chances to outshine his rival, Michelangelo even supplied him with drawings for the Raising of Lazarus, the picture which is now in our National Gallery. Raphael's composition was to be the Transfiguration of Christ on Mount Tabor, which was left unfinished at his death and placed by his bier before his burial. I need not emphasize to what extent these two altar painters had been drawn into the tradition of dramatic evocation, in Sebastiano's case almost at the expense of their religious function, while Raphael had achieved a marvellous balance between the two, with the miraculous scene on the mountain and the despair of the disciples below who, in the absence of St Peter, were unable to cure a boy possessed by a demon.

From the very year of this momentous competition in Rome, which was muted by Raphael's death in 1520, another incident can be documented which illustrates even more sharply the emergence of the new function of the altar painting as a work of art in its own right.[19] It involved the greatest of the Venetian masters, Titian, and one of his principal patrons, Duke Alfonso d'Este of Ferrara. Titian had been commissioned to paint an altar painting for the High Altar of the Church of St Nazaro and St Celso in the North Italian city of Brescia. It is still in that church (Pl. XXXIX). Titian painted in the centre the risen Christ, and on the wings above, in half-length figures, the Annunciation, with the Angel on one side and the Virgin on the other. Below he painted the donor, the papal Legate Bishop Altobello Averoldo who is seen kneeling in prayer under the protection of the two saints to whom the Church is dedicated. One is St Celso, the soldier saint who points to the hope of salvation embodied in the risen Christ. On the wing opposite we see St Sebastian, a saint whose intercession was thought to be particularly powerful against the omnipresent perils of the plague.

Some nine years earlier Titian had also included St Sebastian in an altar painting specifically dedicated as a prayer against the plague (Pl. XL). It shows St Mark, the patron saint of Venice, flanked by the two medical saints, Cosmas and Damian, holding medicine boxes, St Roch who points to the wound which is his

---

[19] The relevant documents are most easily available in Adolfo Venturi, *Storia dell'Arte Italiana*, IX. iii (Milan, 1928), pp. 111–14. I am indebted to Charles Hope for discussing this correspondence with me.

emblem, and St Sebastian having suffered martyrdom tied to a tree as a target for the arrows of his torturers. It goes without saying that here the arrow sticking in the body of the young man is indeed an attribute, a pictographic sign as in Giotto's picture of St Stephen. Nor need I enlarge on the contrast between the way the martyrdom is visualized in the Brescia altar-piece. The change from symbolic rendering to dramatic evocation was not lost on the Venetians. In fact the master's new version and new vision of the event caused an equally dramatic reaction. My final story starts with a letter of December 1520 from Venice to Ferrara addressed to Duke Alfonso by the duke's agent, one Tebaldi.

The agent had been to Titian's studio where he had seen the St Sebastian on an easel (Pl. XLI). He tells his master that all visitors praised it as the best thing Titian had ever done. And to give the duke an idea, he appended a description which is worth quoting in full, for we don't have many such opportunities of hearing what a sixteenth-century layman thought of a particular work of art:

The aforementioned figure is attached to a column with one arm up and the other down and the whole body twists, in such a way that one can see the whole scene before one's eye, for his is shown to suffer in all parts of his person from an arrow which has lodged in the middle of the body. I have no judgement in these matters because I am not a connoisseur of art, but looking at all the features and muscles of the figure it seems to me that it resembles most closely to a real body created by Nature, which only lacks the life.

Nor did Tebaldi hide from us or the duke what conclusions he drew from this display of mastery. He reports that he waited till the crowd had left and then told the painter to send this painting not to Brescia but to the duke, because, as he candidly and significantly put it, 'that painting was thrown away if he gave it to the priest and to Brescia'. The original function, the purpose for which it was demanded and painted, to stand on an altar, was irrelevant in the eyes of the duke's agent. The days of the collector had arrived. It was simply too good for a liturgical role and should be treasured simply as a work of art.

The agent reinforced his plea with a strong economic argument. Titian had been promised 200 ducats for the whole altar, but the duke would pay 60 for the Sebastian alone.

Titian replied that to yield to this request would be an act of robbery, though there are indications that he was not altogether disinclined to commit this act. In the end it was the duke who got cold feet, for he found it diplomatically inadvisable to offend a powerful bishop and legate of the Pope. The painting was left to serve its original function.

And yet, we may feel that it is no accident that it was over this painting that Titian had become involved in a momentous conflict of loyalties. For in a sense it was he who had courted this reaction precisely by the change from symbolism to narrative. He had discarded the last remnant of the medieval heritage for the tradition rediscovered and valued by the Renaissance, the demand for dramatic evocation. His drawings bear witness to the fact that he, like Leonardo before him, was aiming at a masterly solution (Pls. XLII, XLIII).

Art historians have linked these drawings with what was then the most famous statue of antiquity, the Laocoon group recently discovered in Rome in which the beholder is made to witness the agony of the innocent victim and his two sons. The most admired artist of the age, Michelangelo, had taken up the challenge of this group in his images of the dying slaves intended for the tomb of Julius II (Pl. XLIV). It is with such works of intense dramatic evocation that Titian evidently entered into competition, indeed one might say that the dominant demand which the image was now expected to meet, was to stand up to comparison to that canon of excellence. In other words art had created its own context, its own ecological niche, once more, as it had done in the ancient world, and it was this autonomy, this emancipation which led in turn to its survival in a new and hostile climate.

For consider the dates. The year is 1520—the time of the Reformation in Germany and in the Netherlands which was to sweep the images from the altars as merely serving idolatrous heresy. The blocking of this outlet which had provided so many artistic workshops with their livelihood might well have led to the decline and extinction of the image-maker's skill, and there are regions such as Germany where this came close to happening. If it did not happen where the tradition of artistic production was as vigorous as it was in the Netherlands this was due to the fact to which I have alluded, the fact that art had lodged itself in a new ecological niche, the painter's skill was admired for its own sake and the admired specialists in *mimesis* had begun to supply an eager export market.

There is a famous painting of the collection of the Regent of the Netherlands Archduke Leopold Wilhelm (Pl. XLV) on which we recognize many of the treasures now in the Vienna Kunsthistorisches Museum. Quite a number of these paintings were originally intended for altars and private devotions. They had now become Art with a capital A, as it were, they had been cut loose from their roots and flourished in a new environment. And yet the historian remains aware of those roots I have briefly traced in this lecture. Most of them, of course, are easel paintings, a form of art which is peculiar to our Western tradition. If we visit in our mind some of the great national galleries of the world which have extended the chronological span beyond the works collected by the seventeenth-century collector, we discover that the earlier rooms are also devoted to easel paintings, including a number of panels on golden backgrounds dating from the thirteenth, or more probably the fourteenth centuries. They were of course intended to stand on altars in churches or in the home and it is from the moment of their production that we can trace the unbroken evolution of painting in the West to the present day. To quote the words of Otto Demus from his book *Byzantine Art and the West*: 'Had it not been for the transformation of Hellenistic panel painting into Byzantine icon painting, and the transfer of this art form to the West, the chief vehicle of Western pictorial development would not have existed . . .'.[20] Thus, I believe, it is true to say that even the artist today who is, as the saying goes, facing the challenge of the empty canvas or hardboard on his easel, owes his predicament and his joy to the demands made on his predecessors some 700 years ago.

[20] Demus, *Byzantine Art*, p. 205.

Index